Hunters of the Deep

Other Publications:

HOW THINGS WORK
WINGS OF WAR
CREATIVE EVERYDAY COOKING
COLLECTOR'S LIBRARY OF THE UNKNOWN
CLASSICS OF WORLD WAR II
TIME-LIFE LIBRARY OF CURIOUS AND UNUSUAL FACTS
AMERICAN COUNTRY
VOYAGE THROUGH THE UNIVERSE
THE THIRD REICH
THE TIME-LIFE GARDENER'S GUIDE
MYSTERIES OF THE UNKNOWN
TIME FRAME
FIX IT YOURSELF
FITNESS, HEALTH & NUTRITION
SUCCESSFUL PARENTING
HEALTHY HOME COOKING
UNDERSTANDING COMPUTERS
LIBRARY OF NATIONS
THE ENCHANTED WORLD
THE KODAK LIBRARY OF CREATIVE PHOTOGRAPHY
GREAT MEALS IN MINUTES
THE CIVIL WAR
PLANET EARTH
COLLECTOR'S LIBRARY OF THE CIVIL WAR
THE EPIC OF FLIGHT
THE GOOD COOK
WORLD WAR II
HOME REPAIR AND IMPROVEMENT
THE OLD WEST

For information on and a full description of
any of the Time-Life Books series listed above,
please call 1-800-621-7026 or write:
Reader Information
Time-Life Customer Service
P.O. Box C-32068
Richmond, Virginia 23261-2068

THE NEW FACE OF WAR

Hunters of the Deep

BY THE EDITORS OF
TIME-LIFE BOOKS, ALEXANDRIA, VIRGINIA

CONSULTANTS

O. DAVIS BROWN III (Ret.) has worked as a private consultant to the U.S. Navy on various projects, providing technical support for Pacific Fleet exercises and the evaluation of Navy tactics and communications. A former Navy pilot, he has served as a fleet instructor, test pilot, and ship and air wing maintenance manager.

CAREY DORSET is a retired U.S. Navy sonarman experienced in both destroyers and submarines. He is currently employed as a technical consultant to the Naval Sea Systems Command for the Trident submarine trainers and training facilities at Bangor, Washington, and Kings Bay, Georgia.

COMMANDER ANTHONY M. FRANZITTA (Ret.) served on a number of fast attack and ballistic-missile-class submarines. He has also served as a submarine tactics instructor at the U.S. Naval Submarine School and as a battle group antisubmarine warfare instructor at the Atlantic Fleet Tactical Training Group.

CAPTAIN ARTHUR GILMORE is a retired U.S. Navy submarine officer who has had extensive experience with submarine combat systems. He is currently a technical director in the Submarine Systems Operations Center of the Atlantic Research Corporation in Rockville, Maryland.

COMMANDER RICHARD R. PARISIEAU retired from the U.S. Navy in 1980 after serving on thirteen attack and ballistic submarine patrols. He also served as Director of Advanced Submarine Technology at the Office of Naval Research as well as Director of Anti-Submarine/Undersea Warfare within the Naval Intelligence Command.

CONTENTS

Clash in the South Atlantic

Orange life rafts drift on the ocean swell as surviving crew members abandon the mighty *General Belgrano* after a British submarine attack crippled the Argentine cruiser. Bow severed by one torpedo *(far left)* and holed amidships by another, the ship went down some forty-five minutes after being struck, taking hundreds of men with her.

The slender black stem of the submarine's periscope pierced the ocean's placid surface, its monocular lens warily sweeping the desolate seascape. Sixty feet below, at the other end of the telescoping tube, Commander Christopher Wreford-Brown at last glimpsed the prey he had been stalking by sonar for more than twelve hours. In his eyepiece, the dull gray, angular profile of a capital ship—with an oiler alongside and two destroyers patrolling nearby—broke the smooth line of the horizon in the distance.

As captain of the British nuclear-powered attack submarine *Conqueror*, Wreford-Brown had brought his boat 8,000 miles from its home port at Faslane, Scotland, to the southern tip of South America. The *Conqueror* and two sister attack subs, the *Splendid* and the *Spartan*, formed the vanguard of a powerful task force dispatched from Britain to overturn Argentina's occupation of the Falkland Islands.

On April 2, 1982, the military junta ruling Argentina had acted on a longstanding claim of sovereignty over the islands they called the Malvinas. Initially, the generals did not expect the British to fight for these remote, insignificant possessions, despite Prime Minister Margaret Thatcher's unequivocal pledge that the occupation would not stand. But British resolve was firm. Immediately, the government began to assemble a vast armada of carriers, escorts, supply vessels, and troopships to recapture the islands. Taking note of these preparations, widely reported in the press, the Argentines realized that they would have a fight on their hands after all and soon had additional troops and weapons streaming to their exposed island garrison.

London challenged this buildup by declaring a maritime exclusion zone (MEZ). Issued on April 7, the edict announced that any Argentine vessel sailing within 200 miles of the Falklands five days thence would be subject to attack without warning. Until the task force arrived, however, the MEZ could be enforced only by the swift nuclear attack subs, the first of which had sailed the day after the invasion and would be on station by the deadline and well ahead of the surface fleet. Fearing the presence of submarines, Argentina halted the flood of seaborne reinforcements to the islands.

The *Conqueror* arrived at her designated patrol area south of the MEZ on April 30. That evening, a chain of acutely sensitive underwater microphones trailing behind the sub picked up the distinctive noise of surface-ship propellers at long range. After closing on the contact's sonar bearing throughout the night, Wreford-Brown was rewarded with his periscope view shortly after dawn on a day unusually calm for late autumn in these southern latitudes. With the expertise born of fourteen years' submarine service, he had located the cruiser *General Belgrano*, nucleus of one of the Argentine Navy's two principal battle groups. (The other, comprising a carrier and several escorts, lay somewhere north of the Falklands.)

Wreford-Brown's orders precluded him from attacking any Argentine vessel outside the MEZ, as his contact clearly was at the moment. "I went deep and took up a tracking position astern of them," he related. "I had sent the locating report via satellite communications—only a short transmission of a few seconds—and we settled down to an uneventful trail."

The *Belgrano*'s crewmen disconnected and cast off fuel hoses from the oiler, and the ships moved apart. In loose formation with her two accompanying destroyers, the *Bouchard* and the *Piedra Buena*, the *Belgrano* increased speed to thirteen knots on a southeasterly course, unaware of the adversary below matching her every move.

Commissioned in October 1938 as the USS *Phoenix*, this venerable warship had a long and storied past. Her shakedown cruise included a goodwill call in Argentina en route to her new home with the U.S. Pacific Fleet. Three years later, riding at anchor when Japan attacked Pearl Harbor on the morning of December 7, the *Phoenix* blazed away at marauding Japanese planes, which passed her by in pursuit of bigger game along Battleship Row. She came through the air raid intact and went on to pile up an impressive record in World War II, supporting amphibious operations with her big guns and sinking a Japanese battleship in the Battle of Leyte Gulf.

In 1951, the United States sold the *Phoenix* to Argentina. In the following decades, her radar, fire-control system, and electronics suite had been modernized. Though nearly forty-five years old, the cruiser remained a formidable surface combatant. Now she was steaming to war for the first time under a new name and a new flag.

The *Belgrano*'s skipper, Captain Hector Bonzo, had his orders. "Our mission was patrolling the southern zone of the Argentine sea," he explained. Sailing back and forth like a soldier on guard duty, the *Belgrano* could intercept British ships attempting to round Cape Horn from the Pacific and also deter Chile—Argentina's inimical neighbor—from intervening in the Falklands dispute. "We were to go out on an east-west line and then return west," Bonzo said. "We always sailed outside the exclusion zone, never any closer than thirty-five or forty miles."

Rear Admiral John Woodward, commander of the British task force, was, of course, not privy to Argentine strategy. As his surface fleet reached its operational area northeast of the Falklands on May 1, he feared that the *Belgrano* might elude the submarine. To do so, the entire battle group could attempt a high-speed dash across the Burdwood Bank—a large area of water due south of the Falklands too shallow for submarine operations. Or the cruiser and her escorts might split into two or more elements; the *Conqueror* could shadow only one of them. Adding to Woodward's concern, the Argentine aircraft carrier, *25 de Mayo*, had sortied from Puerto Belgrano far to the north—that much was known—but the *Splendid* and the *Spartan*, covering that flank, had not yet found her.

This two-pronged threat preyed on the admiral's mind. His fleet could suffer untold damage from a surprise first strike. Under the circumstances, his best option appeared to be a preemptive attack that would send the Argentines a clear message to keep their distance. "I therefore sought," Woodward revealed, "for the first and only time throughout the campaign, a major change to the rules of engagement to enable *Conqueror* to attack *Belgrano* outside the Exclusion Zone." The *25 de Mayo*—Woodward's preferred target since it posed the greater threat—had so far eluded British eyes. However, the *Belgrano* would serve.

In London, members of the War Cabinet gathered for an impromptu meeting to consider the request. In less than twenty minutes, they deferred to the assessment of the commander on the spot. Although the decision clearly constituted an escalation of the crisis, the government nevertheless hoped that one decisive blow might, in the end, avert a toe-to-toe slugfest in the South Atlantic.

While major decisions were being hashed out 8,000 miles away, the *Conqueror* had patiently followed the *Belgrano* battle group on a route skirting the southern fringe of the Burdwood Bank. In the early-morning hours of May 2, Wreford-Brown watched the three vessels execute an about-face, returning along their outbound course on a reciprocal heading. The weather was deteriorating, but a strong headwind that buffeted the surface ships as they steamed to the west disturbed the *Conqueror* not at all. Shortly after noon, Wreford-Brown received London's authorization to shoot.

Submarine combat is among the rarest forms of warfare. Only once since World War II had an attack submarine fired torpedoes in anger: A Pakistani submarine diminished the Indian Navy by one frigate during a war the two countries fought in 1971. The second such action was about to get under way.

Like any prudent skipper, Wreford-Brown had already worked out in his mind the mechanics of an attack on the cruiser. Eschewing the Mark 24 Tigerfish, a wire-guided, acoustic homing torpedo, "I chose the older, closer-range, Mark 8 torpedo because it has a bigger warhead and therefore a better chance of penetrating the warship's armor plating and antitorpedo bulges—good World War II stuff that isn't in the design of ships these days."

For more than two hours, the *Conqueror* maneuvered to close the range and get into a good firing position. Forced to slow down whenever he came to periscope depth for a look, Wreford-Brown would quickly obtain a visual fix, then go deep and increase speed to make up the distance. "It was tedious rather than operationally difficult," he allowed. "We eventually got them—ourselves on the cruiser's port beam, with the two destroyers on her starboard bow and beam. I think the escorts were mainly thinking of a threat from the north, while we were to the south."

Just before four, in the fading light of late afternoon, the *Conqueror* loosed three torpedoes in quick succession at the *Belgrano*, only 1,400 yards away. Wreford-Brown never expected to score with all of them; he hoped that a three-torpedo spread of the unguided Mark 8s would ensure at least one hit. Inside the sub, the crew heard a loud whoosh as a blast of compressed air launched each torpedo from its tube. The high-pitched whine of their electric-motor-driven props grew fainter as they sped away at forty-five knots; forty-three seconds later came the muffled sound of two explosions.

Through the periscope, Wreford-Brown saw an orange fireball bloom just aft of the cruiser's mainmast, then with the second impact, farther forward, a "spout of water, smoke and debris." A loud cheer resounding through the control room broke his concentration. Tearing himself away from the riveting spectacle in his periscope lens, he was surprised to see every square foot of space in the cramped room packed with sailors watching his every move.

Referring to the simulator in Scotland where he had honed his skills for such a moment, Wreford-Brown commented, "In the Attack Teacher at Faslane, we would have stopped to have a cup of coffee at that stage." But there was no time now to celebrate the victory. He turned his attention to evading the destroyers, which were sure to come looking for him in a vengeful mood.

The *Belgrano* trembled, "just as though it had reared up on a sandbank, coming to a dead stop," said Captain Bonzo later. He had just left his quarters, headed for the bridge, when he heard the explosions. Some of his officers thought the ship was under air attack, but Bonzo guessed at once that she had been torpedoed.

Equipped with delay fuses, the weapons had penetrated well into the ship before detonating. One blast erupted about fifteen yards aft of the cruiser's bow, which disintegrated. Shards of twisted deck plating hung over the void. Fortunately, the watertight bulkheads behind the vanished bow section held, and the powder magazine on the other side of the bulkheads did not explode. Though damaging, this torpedo was probably not fatal to the cruiser.

However, the other torpedo penetrated to the ship's aft machine room, killing all the occupants. Venting upward through the decks directly above the machine room, the blast obliterated two mess halls and a recreation area crowded with sailors just coming off watch. Sheets of flame raced down passageways, trapping and incinerating crewmen before they could flee. Close to 300 men perished in this section of the *Belgrano*. With no other outlet, the explosive force blew a hole twenty yards in diameter through the main deck, producing the fireball Wreford-Brown observed.

This torpedo had also wrecked the firefighting system and generators for the emergency lights, leaving the interior pitch dark and filled with acrid smoke. With the damage control systems out of action, the crew could not check the inrush of seawater through the gaping hole below the waterline. By the time Captain Bonzo made his way to the bridge, the *Belgrano* was already slanting to port. Within five minutes, the list had reached fifteen degrees.

From his perch high in the superstructure, Bonzo could see wounded sailors being raised from the bowels of the ship to the main deck. Many were naked and scorched. "Medics were trying to give morphine or first aid to those suffering burns. It was all very well organized, there was no panic. There were people who went down five or six times to the lower decks to bring up the wounded."

Gun Mechanic Oscar Pardo, a conscript seaman, volunteered for the perilous rescue mission into the ship's murky interior. "She was split open from side to side," he recalled. "There were dead men everywhere, bits of bloody bodies, an arm here, a leg there. We could only help the ones who cried out, because you could hardly see and the smoke was hurting us badly. I couldn't last five minutes."

So that as many men as possible might be saved, Bonzo delayed giving the order to abandon ship until the vessel seemed in danger of capsizing. Then he reluctantly voiced "the most tragic order that a captain can give in his life." Sailors released plastic containers lining the ship's rail that held the life rafts. As a container tumbled overboard, a lanyard secured to the ship pulled it open and activated a CO_2 cartridge to inflate the raft for sailors to clamber aboard.

Lieutenant Juan Meunier had been asleep in his bunk when the *Conqueror*'s torpedoes struck. Hastily dressing, he grabbed his life jacket and hurried up to the main deck. A chief petty officer helped him load crewmen into rafts bobbing in the water next to the port side. With everyone in their area accounted for and safely away, the chief and the lieutenant decided it was time to leave. By now, the port rail was awash and waves were lapping the main deck. "We got into the last raft to go in that section," Meunier recounted. "It may have been the last one to go on the whole ship." But his ordeal had not yet ended. "We kept moving nearer to the bow—or what was left of it," he said. "We eventually had to jump overboard from the life raft because it was being torn apart by the jagged metal there."

Its Falklands mission complete, HMS *Conqueror* glides into its home port at Faslane, Scotland, on July 3, 1982. The piratical Jolly Roger flag *(inset)* is traditionally displayed by Royal Navy submarines after a successful combat cruise. Here it signals the sinking of the Argentine cruiser *General Belgrano* two months earlier.

He swam to another raft, where helping hands pulled him in.

Captain Bonzo remained with his ship until the crew had safely evacuated. He released a few more life raft containers, although there were plenty in the water already. The main deck, slippery with oil, now tilted at a forty-five-degree angle. Fortunately, the fires belowdecks had been extinguished by the flood of seawater before the flames reached topside. Bonzo thought himself alone until he heard the voice of a senior enlisted man behind him. He ordered the sailor to swim for it, but he would not budge until satisfied that his captain was right behind him. "We jumped into the water and reached four life rafts that were about fifty meters away from the ship waiting for us, refusing to leave the side of the ship despite the danger of being dragged down when the ship sank."

Scores of orange rafts dotted the sea. Drenched but grateful to be alive, the survivors watched the *Belgrano*'s demise with morbid fascination. Lieutenant Commander Jorge Schottenheim described the death throes. "The ship settled further on its port side, very slowly but eventually reaching ninety degrees, quite vertical, with the three guns of Number Two turret sticking starkly into the air." Almost gracefully, the cruiser began to slip below the roiling waves, stern first. The truncated bow disappeared, and she was gone.

About six minutes after the explosions that heralded the *Conqueror*'s success, the submarine crew heard the first pings from the Argentine destroyers' sonar. Soon thereafter, the far-off rumble of depth charges reached the submarine. Commander Wreford-Brown decided to withdraw before the probing escorts tracked him down, "a decision which, according to the looks on the faces of the men in the control room, met with everyone's approval."

Almost unbelievably, the *Belgrano*'s destroyer escorts had at that moment no inkling of the tragedy that had befallen their flagship. They had neither seen nor heard the explosions that wracked the cruiser. Distress flares and signal lamps went unnoticed. Generators knocked out and having no backup power for her radios, the *Belgrano* could not alert them to her predicament. They were reacting instead to a supposed torpedo attack against themselves. The crew of the *Bouchard*, hearing a loud thump against the hull of their ship, suspected that a torpedo had struck without detonating. (Subsequent investigation revealed a dent below the waterline; conceivably, the *Conqueror*'s third torpedo, its momentum nearly spent, scored a fluke hit on the destroyer.) In response, the two escorts put on speed and began combing the depths for their phantom assailant as they moved off to the west.

With them went the only hope of immediate rescue for the *Belgrano*'s crew. Many spent two miserable nights in gale force winds and towering seas before search aircraft, dispatched when the destroyers finally noticed the *Belgrano*'s absence and received no reply to their radio queries, located them. Several badly burned sailors died before help arrived, and one fully loaded raft capsized. Its twenty occupants were never found. Captain Bonzo's life raft—with two dead men aboard—was the last to be picked up.

The *Conqueror* returned to the site of the sinking on May 4. Raising his periscope, Wreford-Brown saw "two destroyers and a merchant ship" but made no move to attack. There was no need; after completing rescue operations, the ships returned to port. Though Wreford-Brown did not yet realize it, his actions on May 2 had changed the complexion of the war raging in the South Atlantic.

Learning the fate of the *Belgrano*, the Argentine carrier *25 de Mayo* and her battle group withdrew—having never launched an aircraft or fired a shot—rather than risk destruction at the hands of Britain's nuclear attack submarines. "Their very presence in the area," commented a Spanish analyst after the war, "created a great fear for the Argentinean Navy's command and, after the sinking of the *General Belgrano*, the Argentineans decided to withhold their fleet in mainland's harbors." Rear Admiral Woodward's task force still faced a grueling campaign—land-based fighter-bombers would inflict grievous losses on British ships in the constricted waters around the Falklands—but the seaborne threat to his fleet had evaporated when the *Conqueror*'s torpedoes found their mark.

A Culmination of Generations

A modern attack submarine like the *Conqueror* traces its lineage back to the mid-nineteenth century and the CSS (Confederate States Ship) *Hunley*. On a moonlit night in 1864, nine brave Confederate sailors took this tiny submersible with a hand-cranked propeller out to challenge the Union blockade of Charleston harbor and sank one of the largest ships in the enemy squadron. Though they paid with their lives, the *Hunley*'s crew had dramatically demonstrated the feasibility of submarine warfare.

A half-century later, on the eve of World War I, all the great powers had embraced the concept, though there was no consensus on how best to employ these innovative weapons. Great Britain viewed submarines largely as a novel means of conducting hit-and-run attacks against enemy warships. On the other side of the North Sea, the Imperial German Navy thought in different terms. Forging an underwater blockade around the British Isles to constrict the flow of food and war matériel, German U-boats almost tipped the war in Germany's favor. They came even closer to bringing England to its knees in World War II. Prowling the far reaches of the North and South Atlantic, the submarines devastated convoys in mid-ocean and sank enemy ships off the coasts of a dozen different countries. Had not the Allies radically improved their ability to find and destroy the U-boats, the war would have lasted much longer.

Many German submarines had been sunk when they surfaced to run the diesel engines that recharged the vessels' batteries, used to power electric motors for subsurface propulsion. Even the snorkel—the breathing tube that allowed a U-boat's diesel engines to be run underwater for battery charging—came too late in the war to help the Germans much.

Submarines would not escape the need to approach the surface daily until the postwar development of nuclear reactors. Requiring no oxygen and producing no exhaust, these power plants could run underwater, without a snorkel, for a year or more without refueling. Equipped with such a propulsion system, a submarine could drop out of sight for months on end, as long as provisions lasted. The first such boat, an attack submarine christened the USS *Nautilus*, slipped down the ways in 1954. Since then, these boats have evolved into fast and silent leviathans of the deep.

By their very nature, submarines lend themselves to covert special operations. Furtively creeping up to an enemy's coastline, a sub can reconnoiter shore installations, insert and extract agents, seed a minefield where hostile shipping will pass unawares, or launch cruise missiles while safely submerged. Fitted with ultrasensitive sonar, attack subs can stalk not only surface ships but each other—as well as the huge ballistic-missile submarines with which the United States and the Soviet Union threatened each other for nearly thirty years and that still lurk silently in the depths. ★

Prowlers of the Ocean Depths

Submariners shun the surface—with good reason. On top of the water they are exposed and vulnerable; immersed, they are cloaked with invisibility. Rarely does a sub skipper take his boat above periscope depth while out of port, and he would not risk coming even that close to the surface if there were another way to receive orders by radio and to check his position from time to time. Yet for all the security this cold, lightless domain offers, it also presents challenges.

The ocean depths are unlike any other environment on earth. Tons of water push from all sides, searching for any flaw in a sub's hull. A slight imperfection can give way, allowing the sea to stream into the boat's interior with the force of a fire hose, filling compartments in minutes.

With their sturdy pressure hulls, modern nuclear-powered attack submarines can plumb depths that would have crushed earlier generations of submersibles. And today's subs benefit in speed and agility from the principles of hydrodynamics. Although water is much denser than air, the manner in which a submarine maneuvers through the depths is more analogous to an aircraft in flight than a ship on the ocean's surface, since there is a vertical component to its motion. Movable control surfaces fore and aft deflect the water flowing past the hull, allowing the submarine to go deep or shallow at will. Internal plumbing systems fine-tune buoyancy to keep the boat balanced.

As for navigation, modern technology has given attack subs the means to find their way unerringly through the ocean depths. An on-board inertial navigation system tracks the sub's movements with exquisite accuracy. Radio navigation aids, accessible by raising an antenna above the surface, reveal the boat's location in a trice, either to verify the INS's information or to rectify any error. And while these electronic servants greatly simplify the work of the navigator, they have not eliminated the office. For any number of reasons—from equipment breakdowns to the risk of ascending to periscope depth during combat—the navigator may have to step into the breach with charts, pencils, and rulers to tell the captain where the boat is.

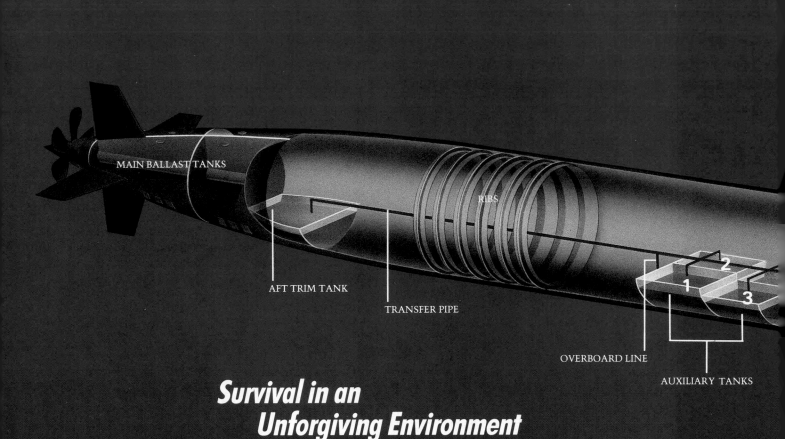

MAIN BALLAST TANKS

RIBS

AFT TRIM TANK

TRANSFER PIPE

OVERBOARD LINE

AUXILIARY TANKS

1 2 3

Survival in an Unforgiving Environment

To operate at great depths, a submarine must be able to withstand the tremendous pressure—as much as 650 pounds per square inch or more—exerted by the water surrounding it. Under this influence, a submarine the size of the Los Angeles-class boat shown here shrinks in volume by nearly 750 cubic feet, the space of a small bedroom in most houses.

Designers have settled on a reinforced cylinder as the most practical, pressure-resistant shape for a submarine. Encased in a streamlined outer hull, this cylinder is called the pressure hull. It is made of reinforced, high-strength steel, one-and-three-quarters inches thick, sufficient to give a Los Angeles-class submarine a maximum operating depth of 1,475 feet. A built-in safety factor is thought to extend the so-called crush depth to more than 2,200 feet. Although the sub may not implode at this level, fittings and seals in the hull could rupture, causing potentially fatal leaks.

To descend to—and safely return from—these depths, the submarine is equipped with both main ballast tanks (MBTs) and trim tanks. With its five MBTs empty, a 6,100-ton Los Angeles-class sub floats handily, even when it is fully loaded for combat. Flooding those tanks adds 800 tons of seawater to the weight of the boat, causing it to submerge.

In most circumstances, the water level in the trim tanks is set for slightly negative buoyancy, that is, so that the vessel will not surface if it slows to a speed where diving planes *(pages 20-21)* are ineffective. The amount of water in the trim system is frequently adjusted to compensate for changes in buoyancy caused by consuming foodstuffs and other expendable supplies and by variations in sea temperature (warmer water makes the boat sink) and salinity (saltier water makes it rise). Trim-tank water is also pumped fore or aft as needed to keep the bow of the sub level with the stern.

The pressure hull *(green)* of a Los Angeles-class sub is buttressed with ribs ten inches wide and spaced less than two feet apart. Although only six ribs are shown here, they extend from one end of the cylinder to the other. At the forward end, a crawlspace leads to a spherical sonar dome in the bow. To regulate buoyancy and keep the sub on an even keel, transfer piping and an overboard line link six trim tanks located inside the pressure hull to the sea and to one another. Like the pressure hull, trim tanks are subjected to much higher pressure on one side than on the other and must be able to withstand the same stresses as the pressure hull. The main ballast tanks at the bow and stern of the boat, fitted with high-pressure air bottles to empty them in an emergency, need not be so strong. Because they are full and open to the sea when the boat is submerged, the pressure inside the MBTs equals the pressure outside.

PRESSURE HULL

MAIN BALLAST TANKS

FORWARD TRIM TANK

AIR BOTTLES

SONAR DOME

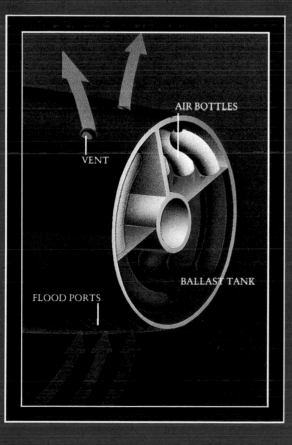

AIR BOTTLES

VENT

BALLAST TANK

FLOOD PORTS

To submerge, vents in the tops of the main ballast tanks are opened, permitting air to escape as seawater rushes in at the bottom through permanently open orifices known as flood ports. Then the vents are closed. To surface normally, the boat first comes to a depth of fifty feet, then the tanks are "blown" with low-pressure air that forces the water out the flood ports. In an emergency, air stored in bottles at a pressure of 4,500 pounds per square inch blasts the water out of the main ballast tanks and keeps external water pressure from collapsing them.

Forces That Help and Hinder

Underwater, a submarine handles more like an airplane than a surface vessel, even banking as it turns. In addition to a rudder for swinging the stern around, it has control surfaces fore and aft—called bow and stern planes—that steer the boat up or down.

Water pressing against control surfaces is the means by which a submarine maneuvers, yet this resistance of water to passing objects is one form of drag that constitutes a major impediment to high speeds underwater. Naval architects strive to reduce drag by streamlining the submarine, as seen in the bullet shape of the outer hull and in the smooth contours of protrusions like the sail and control planes.

Less obvious types of drag also hinder a submerged sub. In pushing forward, the boat creates a wake that generates a low-pressure layer along the hull and sail, causing a type of resistance called form drag.

Most frustrating to engineers is the phenomenon known as skin drag. The thin stratum of water—called the boundary layer—in contact with the sub's hull can erupt into tiny eddies. Their swirling shape disturbs the water's smooth—or laminar—flow, creating almost half the total drag encountered by a submarine.

Why the eddies form baffles scientists, who hope that coatings applied to the hull—or slippery polymers ejected into the boundary layer—will help maintain an uninterrupted laminar flow along the hull.

RUDDER

STERN PLANE

Seated behind the helmsman to his right and the planesman to his left, the diving officer supervises steering, as the helmsman keeps on course and the planesman holds the vessel at the proper depth. Next to him, the chief of the watch sees to trim, as others of the control-room crew carry out various duties.

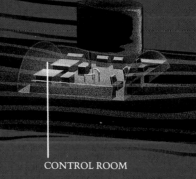

Headed deeper, a submarine has its stern planes rotated slightly up and its bow planes turned partly down. Water pressing on the undersides of the stern planes raises the back of the boat, while pressure against the tops of the bow planes depresses the nose. Dive angle, read from a gauge at the planesman's position, is usually a gentle three to five degrees. The dark bronze color below the sub's water line is antifouling paint applied to inhibit the growth of marine life, which contributes to drag.

CONTROL ROOM

BOW PLANE

TURBULENCE MADE VISIBLE

Made with laser light and specially dyed water, these laboratory photographs illustrate two types of drag. In the top picture, two cylinders standing on end create wakes that are typical of form drag as they are towed through the water. The other two photographs show laminar flow interrupted by eddies that cause skin drag.

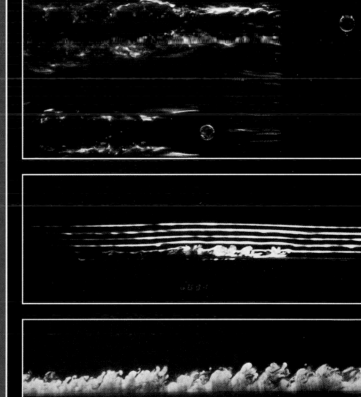

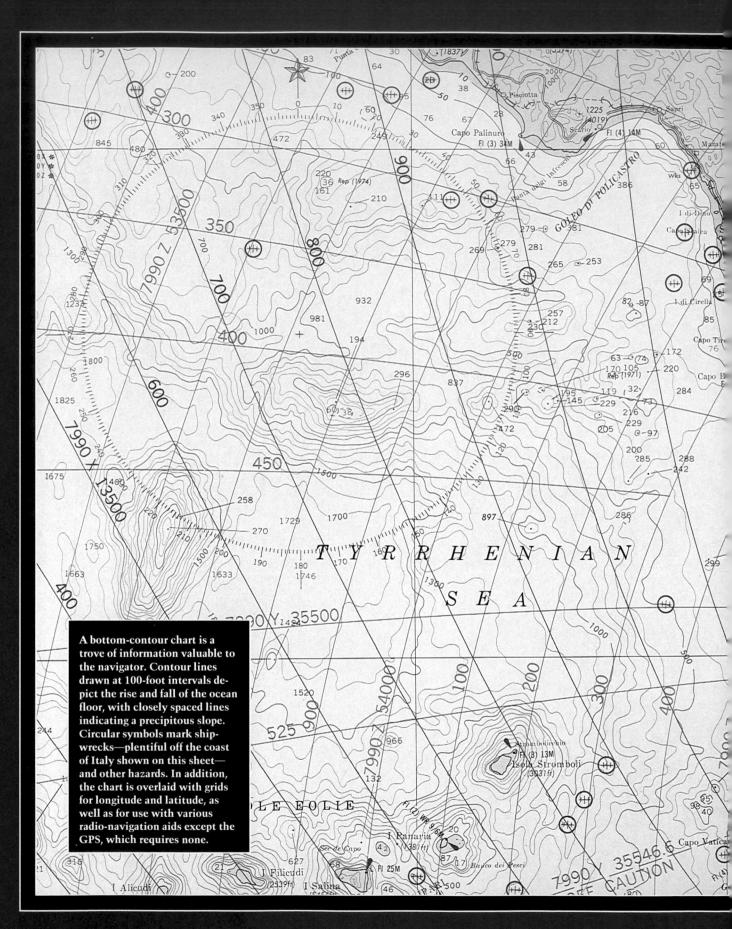

A bottom-contour chart is a trove of information valuable to the navigator. Contour lines drawn at 100-foot intervals depict the rise and fall of the ocean floor, with closely spaced lines indicating a precipitous slope. Circular symbols mark shipwrecks—plentiful off the coast of Italy shown on this sheet—and other hazards. In addition, the chart is overlaid with grids for longitude and latitude, as well as for use with various radio-navigation aids except the GPS, which requires none.

Navigating in the Dark

Submarines navigate primarily with a device called a Ship's Inertial Navigation System (SINS). Similar to equipment installed on many Navy vessels, the self-contained apparatus consists in essence of a computer linked to three gyroscopes and a trio of devices called accelerometers that sense the sub's every movement in all three directions. Taking information from the gyros, the computer continuously calculates the sub's position. As long as the SINS is given an accurate starting point for the boat, the system's precision is phenomenal.

Yet even minor errors accumulate, and after a few weeks, the SINS can have drifted as much as a mile. The solution is to use a navigational aid to give the SINS an updated fix on the boat's position. With a sextant, the sub's navigator can find out where he is by means of celestial navigation techniques. Electronic navigation aids, however, are simpler and quicker. Of these, the Global Positioning System (GPS) is most precise. Consisting of almost a score of satellites, the GPS transmits signals that a special receiver translates into map coordinates with an accuracy of several meters.

All of these methods for updating the SINS require the submarine to ascend to periscope depth, something that is not always possible. Under such circumstances, the navigator has a backup method. He can use a bottom-contour chart of seabed contours and a Fathometer to determine the sub's location by means of bathymetric navigation, as shown on pages 24 and 25.

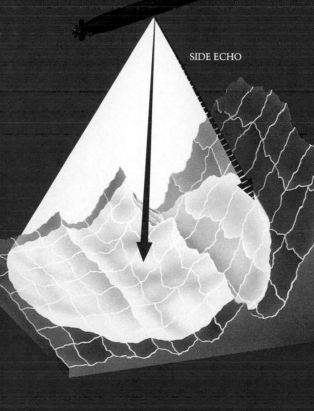

SIDE ECHO

AN EAR TO THE BOTTOM

The Fathometer, a small sonar in the keel of a sub, sends sound pulses toward the bottom in a thirty-degree cone, exaggerated here for clarity. By timing echoes from the pulses, the Fathometer calculates the distance to the bottom. Typically, the first echo to return comes from the point directly below the boat, giving an accurate reading. But in some instances, the earliest echo from a pulse may have been reflected by rising or falling terrain within the cone. Called a side echo, it makes the water seem shallower under the sub than it actually is. Other factors, such as a muddy bottom or layers of plankton that scatter Fathometer pulses, can also cause errors a navigator must contend with when matching Fathometer readings to depths on a chart.

Plotting a Position with Echoes

A navigator usually chooses between two common methods of bathymetric navigation to find a submarine's position: line of soundings or contour advancement. Both depend on taking several Fathometer readings and matching them to bottom contour lines on a chart.

To make either method work, the navigator needs the best approximation possible of his location as a starting point. Usually it comes from the SINS and is refined according to the navigator's log of how much the system has drifted between updates.

Additionally, the bottom below the submarine must vary in height. Since both methods rely on such differences in the ocean's topography to plot a position, they are of little use when a sub is over featureless areas of the ocean floor. For the same reason, the boat must sail across contour lines rather than parallel to them, changing course if necessary to do so. To give readings that correspond to contour lines—which show depth below the ocean surface, not the boat's keel—the Fathometer is set to add the sub's depth to the soundings.

In addition to false readings from the Fathometer, bathymetric navigation is subject to small errors or changes in boat speed and heading, as well as plotting inaccuracies. Nevertheless, above suitable bottom terrain, a skilled navigator can determine a submarine's location within a couple of hundred yards of its actual position.

LINE OF SOUNDINGS

1 Placing a sheet of clear plastic or tracing paper on the bottom-contour chart, the navigator marks his estimated position and draws a line from that point corresponding to the submarine's heading.

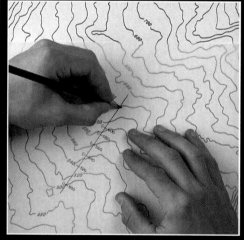

2 As the sub progresses, the Fathometer takes continuous soundings, but the navigator notes on the heading line only returns that match a contour line. He positions them on the heading line according to the time between one reading and the next, the boat's speed, and the scale of the chart.

3 After plotting six or more soundings, the navigator shifts the overlay until the points on the heading line most nearly fall on corresponding contour lines. The last mark becomes the position of the submarine at the time of the most recent sounding.

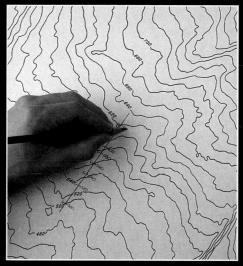

1   In this approach, the navigator draws the heading line directly on the chart. He then plots at least six Fathometer readings as in the line of soundings method shown on the facing page.

2   On an overlay sheet, the navigator draws a straight line and lays it atop the heading line already drawn on the chart. From the chart, he then traces the reference contour line— the one that corresponds to the first sounding noted (500 feet in this example).

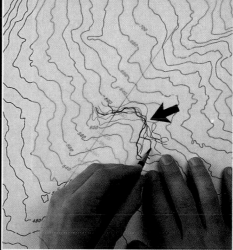

3   Next the navigator adjusts the overlay so that the tracing of the reference contour crosses the heading track at the next sounding mark (520 feet) and traces the corresponding contour line. He continues in this fashion until he has traced all the contour lines that match soundings.

4   Ideally, all six contour tracings would intersect at a single point—the position of the sub when the sixth sounding was recorded. Because of inevitable errors in measurement, however, the tracings offer only a close approximation of the submarine's location.

A Quest for Speed and Stealth

Surging off the bow, a wall of water breaks against the sail of the USS *Helena*, a nuclear-powered Los Angeles-class attack submarine, as she makes a high-speed surface run.

Sometime in 1986, a U.S. spy satellite spotted a Soviet submarine of a new design in the sprawling Komsomolsk shipyard, near the Sea of Japan. When the sleek black vessel—called *Akula*, Russian for "shark"—cast off to begin sea trials, an unidentified American attack sub, submerged offshore, fell in behind. The mission: to compile an acoustic signature of the new ship by recording any sounds it emitted as it slipped almost silently through the depths. Inside the attack sub's dimly lighted sonar room, crew members wearing headsets sat at video terminals, listening for the slightest rhythm that might betray the presence of the quarry. Powerful computers attached to hydrophones trailing the submarine's wake translated the undersea noises overheard by the Navy specialists into beads of light that trickled down the screens, creating a waterfall effect. Eyes trained on the display, the men looked for patterns in the flow signifying the thrum of a passing sub.

In the murky underwater world of submarines, where hunters and the hunted see with sound, the United States had long enjoyed a significant advantage. Unlike furtive American subs, Soviet models typically generated a racket loud enough to be picked up at distances of fifty miles or more with passive sonar techniques—those that listen for noise emanating from an object rather than for echoes of sound pulses beamed toward it. On this occasion, however, the target sub's subtle emanations were virtually indistinguishable from the background noise of the ocean. Indeed, the newcomer seemed nearly as quiet as its top U.S. competitors. Only after sifting the gleanings through complex digital filters in computer banks ashore were Navy sound engineers able to isolate a faint acoustic signature for the Akula.

After analyzing the data, intelligence experts pronounced Soviet submarine technology ten years ahead of where they had expected it to be. Evidently, secrets of silent running—on which U.S. tactical superiority depended—were now known to Moscow.

For nearly four decades, the Americans and the Soviets had vied for dominance of the ocean depths. When the contest began, in the aftermath of World War II, the main function of the submarine was to prey on enemy surface ships. That mission endured, but it was soon overshadowed by new assignments fostered by nuclear technology. The marriage of atomic propulsion with atomic weapons spawned a fearsome new threat—the ballistic-missile submarine, or boomer. Able to lurk in the depths for months on end, an enemy could approach American shores and, without warning, unleash more than a dozen nuclear-tipped lances, each one capable of devastating a small city. Elusive vehicles of mass destruction, boomers came to supplant surface vessels as the principal target of attack subs, which stalked their quarry with homing torpedoes armed either with conventional explosives or with a nuclear warhead.

In the event of nuclear war, attack subs could not hope to detect and destroy all of the strategic subs in the enemy's arsenal before they fired, particularly if the missiles were launched as part of a preemptive first strike. By the 1970s, both the United States and the Soviet Union were deploying submarines with long-range missiles, eliminating the need for those vessels to make extended forays into enemy-patrolled waters. Nonetheless, attack subs were given the mission of searching for the missile launchers near their bases and targeting them once they betrayed their position by firing, if not before.

Besides helping to protect their homelands from assault by intercontinental missiles, attack subs also were assigned vital frontline tasks in the event of a conventional conflict between the superpowers. The Soviets would rely heavily on attack subs carrying short-range guided missiles as well as torpedoes to neutralize American aircraft carriers and to keep vital sea lanes open during a confrontation. For their part, U.S. attack submarines attached to the battle groups would attempt to seek out and silence the predators.

In recent years, the likelihood that hunters of the deep will be called upon to make use of their deadly weapons has diminished in step with the decline of Soviet influence abroad and progress in arms-reduction talks. Barring a ban on ballistic-missile subs, however, the competition to produce quieter hunter-killers may well continue, if only as a hedge against some new and ominous turn in East-West relations. Ultimately, technological advances could deprive submarine tactics of much of their intrigue; subs could

AKULA

The newest class of Soviet attack submarine—its name is the Russian word for "shark"—surprised naval analysts in the West when it appeared in the mid-1980s. The Akula's noise levels were significantly lower than those of its predecessors, an achievement thought to be at least a decade away.

become virtually inaudible to passive sonar, reducing hunting expeditions to matters of luck more than skill, or superior methods of detection could emerge that foil even the quietest designs. But until that time, rival submarine skippers and their crews will continue to play a delicate deep-sea game of hide-and-seek, with little valued more than silence.

Father to the Nuclear Fleet

Coaxing the U.S. Navy into the atomic age was largely the calling of one man—a human dynamo by the name of Hyman Rickover. The son of a tailor who emigrated from Poland early in the century, Rickover outperformed many of his more-privileged classmates at the Naval Academy to gain his ensign's commission in 1922, enhanced his credentials by earning a master's degree in electrical engineering from Columbia University, and proceeded to apply his technical training to the running of the machinery aboard submarines and surface warships with a meticulousness that some found maddening. When he was assistant engineer of the battleship *New Mexico* in the 1930s, for example, Rickover kept the power plant operating at peak efficiency while reducing fuel consumption for heat and light to the point that his shipmates sometimes had to wear their overcoats in the wardroom and grope their way around dimly lighted compartments.

After managing the electrical section of the Navy's Bureau of Ships in Washington, D.C., with a similarly firm hand during the war, he landed the assignment that transformed his career: In 1946, he was dispatched to Oak Ridge, Tennessee, site of the main research facility of the Atomic Energy Commission (AEC)—an organization best known for producing the atomic bomb, but also tasked with developing less apocalyptic uses for nuclear power. There he would investigate possible naval applications of fission reactors—called atomic piles in those days, after the stacks of graphite blocks from which they were assembled.

A Navy maverick who disdained protocol and liked to work through civilian channels, Rickover soon won a powerful admirer among the experts assembled at Oak Ridge—physicist Edward Teller, architect of the hydrogen bomb. Teller referred to the lectures he gave at Oak Ridge as classes for DOPEs—Doctors of Pile Engineering. Captain Rickover, he noted long afterward, was "one DOPE who put his education to good use."

Rickover needed none of Teller's coaching to grasp the far-reaching implications of nuclear power for submarine warfare. At the time, subs under the surface ran on electric motors powered by storage batteries. Electric propulsion systems made minimal noise, but the motors yielded top speeds of little more than fifteen knots, and the batteries needed recharging by diesel-powered generators after an hour or two at full power. Such limited endurance restricted a boat's range, and the moderate pace usually kept it from overtaking large surface ships, the swiftest of which could sustain speeds of thirty-five knots. Instead, a sub had to lie in wait for its prey.

With the almost limitless power of a nuclear reactor, Rickover realized, subs could achieve speeds that would permit them to chase down enemy ships or escort friendly ones. Furthermore, by doing away with the need to recharge batteries, a nuclear propulsion system would, on balance, make a submarine more difficult to detect. Although it would be somewhat noisier than an electrically powered vessel, it would no longer have to risk revealing itself by surfacing or raising a snorkel in order to crank up the diesel engines.

Prospects for such a submarine, however, appeared dim and distant indeed. At one Oak Ridge conference Rickover attended, scientists estimated that two decades of research and development might be needed to build a seagoing nuclear reactor. "Jesus!" Rickover remarked to a young lieutenant seated beside him. "By that time you'll be an admiral and I'll be be pushing up daisies." In his opinion, the major scientific breakthroughs needed to produce a reactor had already been made. What remained to be done was essentially an engineering chore—and Rickover was convinced that the task could be greatly accelerated given a sufficient investment of resources. On his return to Washington, he drafted a report for Rear Admiral Earle Mills, assistant chief of the Bureau of Ships, confidently asserting that the Navy could field a nuclear-powered vessel in no more than eight years—and perhaps as few as five.

Rickover's enthusiasm for the nuclear option proved contagious, influencing officers at the highest level. In December 1947, Chief of Naval Operations Chester Nimitz, who had commanded subs in the Atlantic before leading the U.S. Pacific Fleet to victory during World War II, dispatched a memo to Secretary of the Navy John Sullivan insisting that the future security of the service required the development of "a true submarine, that is, one that can operate submerged for very long periods of time and is able to make high submerged speeds." Only a nuclear-powered vessel promised to meet both those conditions, the memo concluded. Lending urgency to the argument were projections that the Soviets could match or exceed U.S. production of conventional submarines within the next decade. If the Navy's sub fleet hoped to maintain the edge it had gained during the war, it would have to make a technological leap.

Swayed by Nimitz, top officials in Washington soon approved the project, although the terms of the deal left the Navy uneasy. In an awkward bureaucratic compromise, the AEC—already burdened with heavy responsibilities for nuclear development and weapons production—was assigned to build the reactor while the Navy was tasked with designing a new sub for it. Concerned that the AEC would act too slowly, the Navy arranged for the feisty Rickover to be appointed to the commission, trusting in him to prod the reactor effort along. Once ensconced, Rickover outmaneuvered bureau

Atomic Power for Submarines

A nuclear sub's propulsion system fills more than a third of the space within the pressure hull. Between shielded bulkheads, the reactor generates tremendous heat and radioactivity, which raise the temperature of the reactor water *(red)*. Maintained at a constant pressure, this water—many Soviet designs use a molten metal—passes its heat to water in a boiler that provides steam for running the submarine *(green)*. Piped into the engine room, the steam spins two turbines—one to generate electricity, the other to turn the propeller. After being cooled by seawater in a condenser, it returns to the boiler and repeats the cycle.

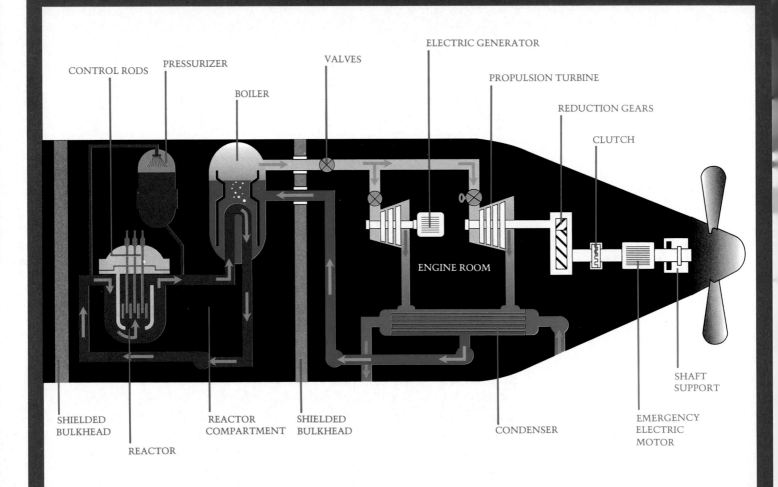

CONTROL RODS
PRESSURIZER
BOILER
VALVES
ELECTRIC GENERATOR
PROPULSION TURBINE
REDUCTION GEARS
CLUTCH

ENGINE ROOM

SHAFT SUPPORT

SHIELDED BULKHEAD
REACTOR COMPARTMENT
SHIELDED BULKHEAD
CONDENSER
EMERGENCY ELECTRIC MOTOR
REACTOR

chiefs and made himself the de facto head of the nuclear propulsion program, ensuring that it would proceed at an urgent pace.

While doing his utmost to urge development forward, Rickover made safety a top priority, working hard to minimize the chance of a nuclear accident aboard a sub. Loss of life to a failed reactor, he felt, would surely turn the public against naval applications of nuclear technology and sink the program. To prevent such an outcome, he shied away from even the most basic automation of reactor controls—a policy that necessitated a big submarine to house a large crew but allowed for relative simplicity, ruggedness, and reliability in the ship's nuclear plant and mechanical systems.

The main challenge facing the reactor design team, of course, was to find a way of exploiting the awesome energy of atomic fission without irradiating the crew. Somehow, radioactivity from the fuel in the core of the reactor would have to be contained even as the brutal heat generated by the pile was put to work. In an elegant solution, ultimately adopted by the commercial nuclear power industry, the team devised a reactor with two loops. The primary loop pumped coolant through the radioactive core and circulated the scalding effluent through pipes; those in turn transferred the acquired heat—but not the radioactivity—to a secondary loop filled with water, generating steam to turn twin propellers *(left)*.

Engineers differed as to the choice of a coolant for the primary loop. Water was one option, but once it passed through the hot core it would have to be pressurized to keep it from boiling away, adding more machinery, weight, and noise to the loop. Another possibility was a coolant with a much higher boiling point, such as liquid sodium. This approach would eliminate the need for a pressurizer.

Intrigued by both proposals, Rickover elected to produce two nuclear sub prototypes: the *Nautilus*, laid down in 1952 to try out the pressurized-water reactor, and the *Seawolf*, authorized that same year to accommodate a liquid-sodium plant. During subsequent dockside tests, the reactor for the *Seawolf* developed alarming leaks, evidently a result of corrosion as the sodium coursed through the pipes. Engineers worked to correct the problem and thought they had it under control, but Rickover had lost faith in the liquid-sodium plant and ordered it replaced with the water-cooled type. Some criticized the decision as premature, but Rickover felt confident that the *Nautilus* and her reactor were sound and saw no need to pursue alternatives. From the start, he had kept a fatherly

33

eye on the *Nautilus* as she took shape at a shipyard in Groton, Connecticut. Scrambling around a life-size mock-up of the vessel, he had made certain that every valve and switch was properly positioned for safe and convenient operation.

With space for a crew of 100 and the bulky reactor plant, the prototype that emerged displaced a hefty 3,500 tons, roughly twice as much as a big World War II-vintage submarine. Although built to demonstrate cruising range rather than firepower, she was a fully equipped ship of the line armed with six torpedo tubes.

When the *Nautilus* slipped out of her berth for trials in January 1954—fulfilling Rickover's stated goal of producing a nuclear sub within eight years—she managed a top speed underwater of nearly twenty-three knots, some eight knots faster than the swifter diesel-electric boats of the day. Far more impressive, however, was the vessel's endurance. The most advanced submarines before her could not exceed fifteen knots below the waves for more than ninety minutes. Soon after her initial trials, the *Nautilus* spanned the 1,400 miles from New London, Connecticut, to Key West, Florida, without surfacing, covering the distance in less than three days at an average speed of nearly twenty knots. Later, in the summer of 1958, she topped all her previous exploits by diving under the Arctic icecap near the Bering Strait, passing directly below the North Pole, and surfacing off the coast of Greenland—an epic passage of 1,830 miles in four days.

Looking ahead to possible peaceful uses of the technology, President Dwight D. Eisenhower praised the *Nautilus*'s polar journey as opening "a new commercial seaway between the major oceans" for prospective nuclear-powered cargo submarines. The only freight nuclear subs would carry in the near term, however, would be fearsome consignments of ballistic missiles. By early 1959, shipyard workers at Groton were putting the finishing touches on America's first boomer, the *George Washington*, a boat with the capacity to fire her weapons while she was submerged. And the Soviets were not too far behind. They had launched their first nuclear attack sub just a few months before the *Nautilus*'s cruise under the pole, and they were already converting some of their larger diesel-powered attack boats to missile launchers. Few in Washington doubted that this was merely a stopgap measure and that the Soviet Navy would soon begin building nuclear-powered boomers similar to the *George Washington*.

The prospect of such a threat lurking hundreds of feet below the waves for months on end forced a searching reappraisal of antisubmarine warfare (ASW), traditionally conducted by ships and planes that searched for subs on or just beneath the surface and targeted them with depth charges or torpedoes. However, ASW forces could no longer count on catching an enemy submarine as the waves lapped its sides. Nor could they afford to wait for a sub to signal its position by firing, for now the sub's target might be not a single merchantman but a whole city.

Ships and planes patrolling a large area without a hint as to the enemy's location—whether it be a periscope sighting or a recently reported attack—could have little hope of intercepting a lurking sub. Planes and helicopters carried sonar units that they lowered on cables and dunked in the ocean, but the small size of the devices limited their power and sensitivity. Ships could carry powerful sonars in their bows, but the thrashing of their own propellers would warn off an enemy sub long before a ship could detect it. Consequently, nuclear attack subs would have to do the lion's share of the hunting for boomers and other attack subs, and success would depend largely on the predator's ability to be stealthier than its prey.

For Rickover, now a rear admiral and firmly established as overlord of the Navy's nuclear program, the challenge was as great as any he had faced in his career. Successors to the *Nautilus* would have to run faster to fulfill the enduring mission of pursuing enemy surface ships and escorting friendly ones in a conventional conflict. But the primary task of attack subs would be to pinpoint enemy boomers and hunters before being found by them, and in that contest sheer speed would count for little if the very machinery that increased power made a submarine noisier and easier for the other side to detect.

In Pursuit of a Quiet Killer

The need to balance speed against stealth became painfully apparent to the Navy in 1959, when a new variety of attack sub known as the Skipjack class made its debut. In the works for several years, the Skipjacks exhibited unprecedented speed, combining, as they did, the awesome range of nuclear power with the swiftness and agility conferred by close attention to hydrodynamic principles. In

Returning in triumph from her unprecedented voyage beneath the polar icecap in August of 1958, the USS *Nautilus* sails into New York harbor with an escort of tugs. By and large, the 107 crewmen were nonchalant about their accomplishment. "We didn't know we were heroes," said one self-effacing sailor, "until the president told us."

A model of the revolutionary submarine *Albacore* sits at the mouth of a wind tunnel during streamlining tests at a U.S. Navy facility near Washington, D.C., in 1948. Engineers often preferred air over water for this kind of work in part because making observations and taking photographs of an experiment were simpler in a wind tunnel than in a tank of water. A simple mathematical formula compensates for the lesser density of air.

designing the *Nautilus*, engineers had paid scant attention to streamlining—a shortcoming the Navy tacitly acknowledged by building only a few operational subs of her type, officially designated the Skate class. Like World War II submarines, built to run much of the time on the surface, the *Nautilus* and her classmates resembled surface ships in their contours, with sharp prows and angular hulls that made for substantial water resistance, or drag, when the vessels submerged.

Just how much the shape of a sub affected its deep-sea performance had been forcefully demonstrated a few years earlier by an experimental diesel-electric boat dubbed the *Albacore*. Named for one of the great swimmers of the deep, the *Albacore* glided efficiently through water. Her graceful contours—rounded at the fore and tapering toward the stern like an elongated teardrop—greatly reduced the so-called form drag associated with blocky designs. At the same time, her designers reduced a second source of resistance known as skin drag by minimizing the hull's total surface area and thus lessening friction with the water—an improvement made possible by a fundamental change in construction.

Submarine designers had long resorted to double hulls, with the angular external hull encompassing a cylindrical inner pressure hull, whose rounded shape helped it to withstand the pressure of the sea at great depths. Between the inner and outer hulls were the sub's ballast tanks. Filling these tanks forced the boat down, while expelling the ballast water made the craft buoyant and impelled it upward. The problem with double hulls was that in order to have sufficient space within the pressure hull to accommodate men and machinery, the outer hull had to be bigger still, increasing skin drag and slowing the sub.

Forsaking tradition, the designers of the *Albacore* opted for a single sturdy hull of welded steel with ballast tanks fore and aft. This arrangement—applied eventually to all U.S. nuclear subs—proved nearly as resistant to pressure as a double-hull sub with the same interior space and gave the sea less to pull on, augmenting the benefits of the *Albacore*'s clean lines. In one test, the racy little craft managed thirty-three knots—a remarkable performance for a submarine whose electric propulsion system was only marginally superior to those of World War II.

Work on the *Nautilus* was far advanced by the time the *Albacore* revolutionized the Navy's thinking, but the test craft's hydrodynamic lessons were duly applied to the next nuclear generation—the Skipjacks. Those attack subs carried a more powerful version of the *Nautilus's* original pressurized-water reactor, which turned a single massive propeller instead of two smaller ones, thus eliminating the bulk of the extra shaft and turbine. With their leaner power trains and more potent reactors, the Skipjacks managed speeds in excess of thirty knots. Although sub commanders admired the graceful Skipjacks for their agility and pep, they found them unacceptably noisy—a defect rendered all the more disturbing by recent improvements in sonar detection.

This problem did not catch Rickover and his engineers entirely by surprise. They knew when they embarked on the nuclear program that it would be difficult if not impossible to match the silence of electric propulsion. Yet the Skipjacks' loud signature, combined with the new emphasis on stealth as the Navy prepared to pit subs against subs, forced marine engineers to reconsider every facet of submarine design with an eye toward noise abatement.

Chief among the acoustic culprits, they realized, was the atomic plant. Throbbing pumps shuttled water through the primary and secondary loops, adding their voices to the hiss of the steam generator and the whine of the turbine, spinning at a furious 6,000 to 7,000 rpm. To compound the racket, reduction gears interacted constantly to cut that rate to about 200 to 280 rpm, the maximum the propeller could handle without simply churning at the water like an egg beater. Even at slow speeds, the rotating propeller produced a noise of its own, comparable to the hum of a large fan. And at medium to high speeds, depending on the depth, the propeller could cavitate—or generate bubbles that collapsed with a sharp popping sound. The best defense against this was an experienced commander, who knew just how fast he could go at a given depth without causing cavitation. Still, naval engineers were busily investigating ways to raise the speed threshold at which propellers began to spawn these bubbles and to reduce other prop noise through improved design and manufacturing techniques. A final source of unwanted sound was the flow of water around the hull. This problem had been greatly alleviated by the recent efforts at streamlining, but there remained room for minor improvements.

In fashioning a stealthy successor to the Skipjacks, designers concentrated on the overriding problem of propulsion-system noise. To muffle the booming power train, they mounted it on a so-called raft—a padded frame, isolated from the hull, that acted as a shock absorber and helped keep the noise from reaching the outer shell of the boat and radiating through the water. This buffer added bulk and necessitated a larger hull, increasing skin drag and resulting in the sacrifice of a few knots of speed relative to the Skipjack. But at the same time, the innovation promised to boost the new vessel's top "quiet speed"—the pace at which a sub could hunt for others without spoiling passive sonar reception with its own noise—and skippers were more than willing to accept the trade-off.

Unfortunately, the first member of this furtive new class, the *Thresher*, came to grief while demonstrating another of its capabilities—deep diving. Boasting a new hull of high-tensile steel with a tubular midsection, she was designed to descend to an unprecedented 1,300 feet—about 300 feet deeper than the Skipjacks. Increased depth made the sub that much harder for surface ships or aircraft to detect and target; more important, it offered a skipper extra seconds in which to save his vessel if human or mechanical error caused a fast-moving sub to dive at too sharp an angle.

On April 10, 1963, the *Thresher* was operating near her 1,300-foot limit during a trial run when she developed a leak that some analysts later attributed to faulty welding in the steel hull. The water caused a short circuit that shut down the reactor, leaving the *Thresher* without power to drive to the surface. When the captain tried to blow ballast, compressed-air valves used to eject water from the tanks malfunctioned. With no way to increase buoyancy, the vessel sank. All 129 men aboard perished. A painstaking investigation of the incident led to tighter quality control during hull construction and the redesign of the ballast system to ensure that the *Thresher*'s sister subs, newly designated the Permit class, could descend to maximum depth without risking a similar fate.

The Permits and their worthy successors—the marginally larger, quieter, and slower Sturgeons—gave the Navy's fleet of attack subs a decided advantage over the Soviets in stealth. Yet Rickover and his aides feared that the Soviets might compensate by building quicker subs that could pursue aircraft carriers and other fast surface ships. If so, the submarine fleet would be hard-pressed to provide effective escorts for naval task forces. The thirty-knot limit of the Permits, for example, was a few knots below the top pace of most U.S. aircraft carriers. And when a carrier task force was responding to a summons in times of crisis, speed was of the essence.

A Visual Record of Sound

Sonar contacts appear on this "waterfall" display as bright lines against the dimmer green of background noise. Set to show sounds at any frequency the sonar can hear, the display is divided into three rows and four columns. In this mode of operation, one of several, each column represents an angle—from 3 degrees above the centerline of the boat to 45 degrees below it—in which the sonar's 360-degree fan of sensitivity is directed. Above each column, a pointer indicates the sub's heading (266 degrees, or nearly due west) on a scale of compass directions; a bright arrow called a cursor, positioned by the operator, shows direction to a contact (333 degrees, or north northwest). Rows represent three contact histories for each of the four angles. When it is first detected, a sound source appears simultaneously at the top of all three, but it moves toward the bottom of each at a different rate. It crosses the top band in just ten seconds, alerting an operator to a new contact or change in an old one. If the contact remains within range, its line will trickle down both the ten-minute band and the forty-minute band, allowing the sonarman to track it over time. The most prominent contacts shown here (A and B) are ones of long standing, a vertical line indicating that there has been no change in the direction to the sound source. Short, diagonal lines represent two contacts briefly heard crossing the sonar's field of view, one from left to right within the last minute (C) and the other in the opposite direction some six minutes earlier (D).

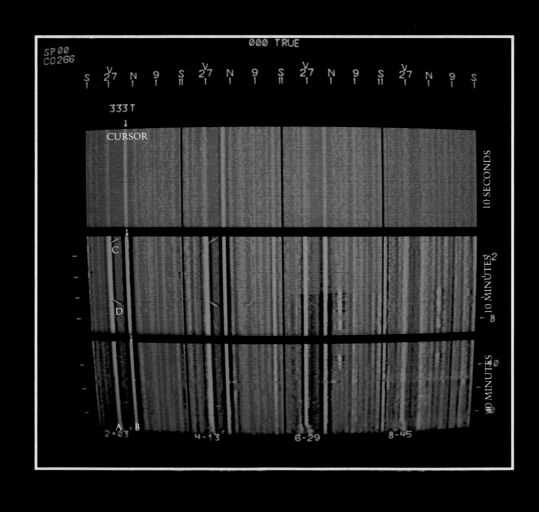

In the mid-1960s, the Navy began to lobby for a new attack sub that would be as fast as the Skipjacks but considerably quieter, one whose duties would include escorting carrier task forces and watching for Soviet hunters. Secretary of Defense Robert McNamara, burdened by the increasing costs of the Vietnam War, questioned the need for another expensive development program. In late 1967, he released a Defense Department study estimating that Soviet submarine technology had been so outpaced by American know-how that in a war, the enemy would lose twenty-five subs for every U.S. sub they destroyed. McNamara proposed adding a few Sturgeon-class subs to the U.S. force and shelving the proposed fast attack sub. Aghast, Rickover and company prepared to take their case to Congress, armed with satellite photos of busy Russian shipyards and new figures showing heavy Soviet expenditures for new submarine development.

Shortly before the hearings were to begin in the Senate, the Navy received a bit of serendipitous support from the opposition. On January 5, 1968, a ship with a U.S. task force detected a Soviet nuclear attack sub trailing the nuclear-powered aircraft carrier *Enterprise* off the coast of California. In a routine evasive maneuver, the *Enterprise* accelerated, expecting the sub to drop away. When she did not, the carrier pushed to thirty-one knots, near her top speed. To the surprise of the *Enterprise*'s captain, the pesky sub all but matched her stride, sprinting to thirty knots before breaking off pursuit. Plainly, the Soviets posed a threat.

A Hasty Reply to Uncle Sam

Analysis of the submarine's acoustic signature revealed that the surprisingly speedy Soviet boat was a member of the decade-old November class. Like ambitious parents pushing their offspring forward, the Soviets had hurried their first nuclear attack sub into the water in late 1958, four years after the United States commissioned the *Nautilus*. Dubbed the *Leninsky Komsomol*, this original November sub—as Western analysts had christened it—was some forty feet longer and displaced 800 more tons than the *Nautilus*. With two pressurized-water reactors supplying 35,000 horsepower to twin propellers and a more streamlined exterior than the *Nautilus*, November subs could overcome the skin drag of their relatively large hulls and accelerate to impressive speeds, as the *Enterprise* task force later learned. That the Americans did not know for ten years the true speed of these subs reflected the difficulty of gathering intelligence about submarines during the 1960s.

Soviet designers were aware of the benefits of single-hull configurations but adhered to the traditional double hull because it was somewhat simpler to manufacture and gave greater protection against torpedoes with high-explosive warheads, which might breach one section of the compartmentalized external hull without necessarily disabling the entire ballast system or rupturing the vital pressure hull. This emphasis on survivability reflected the defensive-mindedness of the Soviet Navy, whose commanders assumed that the larger U.S. fleet would take the initiative in any conflict and who, as a result, sought vessels that could withstand attack and mount selective counterstrikes.

Judged in terms of speed, service depth, and the capacity to withstand shocks, the November-class submarines stood up well to American competition. In other respects, however, the design was primitive, incorporating flaws that would have disqualified it by Western standards. Not only were the November subs noisy—far noisier, indeed, than the racy Skipjack class that emerged around the same time—but living spaces for their crews were miserably cramped, and the ships were plagued with power plant problems. The average time between failures, for example, was less than 1,200 hours—at a time when the U.S. Navy counted on as much as 24,000 hours of dependable operation from their reactors and associated machinery. Although details remain sketchy, design flaws and slipshod maintenance apparently led to a number of accidents aboard November-class subs in the 1960s and the 1970s, some of them involving releases of radiation that are still exacting a toll among Soviet Navy veterans.

Such safety problems reflected both the haste of the Kremlin to catch up with the United States and the fact that mishaps were largely kept secret, precluding public inquiries that might have forced higher standards. As a result of the tight secrecy, little was known to the U.S. Navy of the class's infirmities—except that the vessels were quite noisy—at the time that one of the subs chased after the *Enterprise.* Ironically, ignorance of the other flaws only made Washington more determined to meet Moscow's challenge. Classified reports of the *Enterprise* incident, released to Congress, helped the Navy win its case for fleeter attack submarines to be known as the Los Angeles class.

Equipped with more potent nuclear reactors that turned even larger turbines, these new subs could maintain an underwater speed in the low thirties—a seemingly small gain over the Permits and Sturgeons, but one that allowed them to keep up with aircraft carriers traveling at full steam. More to the point, they were better equipped to hunt enemy subs, not only because they carried superior sonar but because their own acoustic signature proved fiendishly difficult for opponents to make out. Their larger reactors and turbines notwithstanding, Los Angeles subs emerged as the quietest nuclear predators, Soviet or American, ever to take to the water.

Bent on smothering every stray decibel of sound, engineers had scrutinized each grommet, pipe, hinge, and latch in the *Los Angeles* for potential squeaks and rattles. The bulkheads were lined with noise-deadening foam. Every piece of machinery in the ship that might conceivably make noise was precision manufactured, from air conditioners to the power plant. Isolated on a raft similar to the one pioneered in the *Thresher,* the propulsion system was reportedly so well balanced that a coin stood on edge atop the turbine housing would not be toppled by vibration from the machinery.

In addition, the sub's propeller was carefully configured to minimize noise. Part of the improvement could be attributed to the propeller's greater diameter, which allowed it to thrust the vessel forward at lower rpm than smaller predecessors—reducing the sound much as a slow-revolving ceiling fan moves an equal volume of air more quietly than a higher-rpm table fan. In addition, for the first time, engineers employed a seven-bladed propeller, in keeping

On June 26, 1989, a Soviet Echo II-class nuclear attack sub operating off the Norwegian coast suffered a rupture in the cooling unit of one of its two reactors. Shutting down the reactor to prevent it from overheating—and perhaps melting—the crew quickly brought the boat to the surface, where it managed to limp along on auxiliary diesel power venting steam from the cooling system *(left).*

Such a failure aboard a submarine tends to be more perilous than similar incidents aboard surface vessels. The Soviets were fortunate that the Echo II mishap released no radioactivity and injured none of the ninety-man crew.

Other submariners—both Soviet and American—have not been so fortunate. During the Cold War, a total of five Soviet boats were lost at sea, and two U.S. nuclear submarines—the USS *Thresher* in 1963 and the USS *Scorpion* in 1968—went down with all hands aboard.

The lack of survivors in most sub disasters leaves investigators groping for explanations. But in one instance—another Soviet mishap that occurred only three months before the Echo II's near tragedy—there were eyewitnesses to tell the tale. Furthermore, the advent of glasnost in the Soviet Union lifted the veil of secrecy that usually surrounds Soviet fiascos, and the story became public.

Captain Yevgeniy Vanin and his sixty-six-man crew of the Mike-class attack boat—a one-of-a-kind prototype—had completed the sub's first operational deployment and were returning home. The Mike was at a depth of 150 feet on the morning of April 7, 1989, when an alarm flashed in the control room, indicating that the temperature in the sternmost compartment, number 7, had risen to 158 degrees Fahrenheit. An electrical short circuit had started a fire that killed the watch stander and was now burning out of control.

Captain Vanin sent Warrant Officer Vladimir Kolotilin into compartment 6 to activate the remote fire extinguishing system—which uses freon to suffocate the flames—in number 7. Unfortunately, the fire had melted plastic seals on a high-pressure air line running through the compartment. The rush of escaping air fueled the blaze and nullified the effects of the freon. The temperature rose to almost 1,500 degrees, incinerating the gasket around the bulkhead to compartment 6.

Kolotilin picked up the intercom and reported, "I see smoke seepage." Then the intense heat ignited hydraulic fluid in the turbine generator in compartment 6. The control room heard Kolotilin's last words, "It's spurting like a flamethrower."

At that point, a power surge swept through the boat's damaged electrical system, creating more havoc. "Cables and panels were on fire in battle stations," a survivor recalled, "We saw them explode, and guys began ripping them out with their bare hands to try to stop the flames somehow." Luckily, a heroic member of the crew managed to shut down the reactor.

With the situation now critical, Captain Vanin surfaced the Mike in choppy seas and sent a coded distress call to fleet headquarters. As the submarine filled with smoke, damage-control teams donned masks connected by flexible hoses to a compressed-air system. But the fire in compartment 7 had

ruptured these pipes, too. Several firefighters collapsed from carbon monoxide poisoning before the problem was discovered. Those still on their feet switched to portable oxygen masks and hauled their unconscious shipmates up on deck.

Explosions were heard from both 6 and 7. Already weakened by the extreme heat, the pressure hull ruptured, flooding the two stern compartments. On the bridge, Captain Vanin knew the Mike was beyond saving and gave the order to abandon ship.

As crewmen struggled on the pitching deck with a twenty-five-man raft, a gust of wind blew it into the water upside down. Sailors jumped into the frigid sea after it. Captain Vanin, meanwhile, had returned belowdecks to collect stragglers when the Mike abruptly stood up on her stern and sank, her bow plane striking and killing a swimming sailor.

The captain and four other trapped crewmen reached the sub's escape capsule, which is built into the sail. The release mechanism, however, jammed, and the five occupants rode the Mike on its terrifying descent to the bottom 5,000 feet below. Either an internal explosion or the force of the sub striking the seabed freed the capsule, which then rocketed upward.

As it popped to the surface, the hatch blew off, propelling one sailor into the sea, where he drowned. Another crewman, who was thrown halfway out, managed to swim away before the capsule filled with water and sank with the other three men, including Captain Vanin. Soviet rescue ships picked up survivors from the overturned raft, but the accident claimed the lives of forty-two submariners.

with a recent discovery—already being put to use by the Soviets—that sounds made by an even number of blades tend to reinforce one another. Finally, each blade was honed to exacting tolerances using computer-controlled milling equipment to ensure that neither surface roughness nor variations in blade shape could cause rackety choppiness in the wake.

To curtail noises generated by the hull passing through water, external connections, cleats, and hatches were designed to fold into the deck, and the skin was coated with a sound-absorbent finish. In the mechanical equivalent of biofeedback, sensors implanted in the hull enabled crewmen aboard the *Los Angeles* to monitor the vessel's acoustic pulse for any increase in volume that might signal a mechanical flaw requiring correction. Generally, there was not much to hear: At cruising speed, the *Los Angeles* beamed barely 0.01 watts of acoustic power—comparable to the sound of a car zipping along the freeway.

A splendid synthesis of thoughtful engineering and rigorous production standards, the Los Angeles class became the mainstay of the American sub fleet. Since the first operational model left her berth in 1976, more than fifty have been completed and another dozen or so ordered. The class has won the admiration of commanders as much for its furtiveness as for its sustained high speeds.

Moscow Ups the Ante

Although the November-class submarine gave a crucial push forward to the Los Angeles-class program, other Soviet boats that would pose a greater threat to U.S. superiority beneath the waves were in the offing. As early as 1956, while work on the November class was proceeding, the Soviets had begun to contemplate a submarine of radical design that would shatter the speed and depth thresholds inherent in existing technology. Perhaps influenced by feverish speculation in the Western press that American subs would soon reach speeds of forty-five knots and dive to 3,000 feet, Admiral Sergei Gorshkov, commander in chief of the Soviet Navy, sought support from Premier Nikita Khrushchev for a crash program that would match or exceed such performance standards. A technophile obsessed with the idea of upstaging American initiatives, Khrushchev enthusiastically endorsed the project.

Undergoing maintenance in a floating dry dock, the USS *Salt Lake City* displays her smooth snout, which contains the sub's main sonar array. Made of fiberglass, which conducts sound waves better than steel, the bow section is perforated with tiny holes so that the sonar inside is always immersed in water.

Achieving major advances in speed and depth meant rethinking the design of the nuclear power plant and the surrounding hull. Aware that the Americans had experimented with liquid sodium as a reactor coolant in place of pressurized water, the Soviets looked closely at other options and settled on a mixture of lead and bismuth. With a much higher boiling point than water, the liquid lead-bismuth coolant would require no pressurizer to keep it from boiling. Better able than water to absorb heat, the coolant was needed in smaller volume, saving space while at the same time generating more steam pressure for greater shaft power. In addition, crew space was to be minimized by the liberal use of automated controls throughout the sub, reducing the size of the boat's complement to perhaps forty or fifty men—less than one half the number carried by U.S. attack subs. The result would be a significantly smaller sub with superior agility and speed underwater.

As if the switch to automation and a revolutionary power plant were not challenges enough, the designers of this high-performance boat elected to fabricate the vessel's pressure hull from an experimental titanium alloy with half the density and nearly twice the strength of steel. This would give the new ship the ability to withstand deep dives—and an added measure of protection against the blast of a torpedo. Soviet scientists had been exploring new applications for titanium, which was in plentiful supply in the USSR, but the gap between theory and practice proved enormous. More than a decade went by as technicians grappled with new welding techniques to join hull sections fashioned from the titanium alloy.

The first of the wonder subs, designated the Alfa class, at last debuted in 1969, only to retreat to her berth after developing severe cracks during sea trials. All attempts to repair her failed, and she was cut up for study. Years of experimentation ensued. Not until the late 1970s did the Soviets dare entrust their pet project—known wryly in Moscow as the Golden Fish for the many billions of rubles poured into the endeavor—once more to the water.

This time, the Alfa weathered her trials, and reports of her remarkable performance began to filter back to the West. Her two lead-bismuth reactors, driving a single propeller, were said to supply 47,000 horsepower—30 percent more than furnished by the propulsion system aboard a Los Angeles-class sub, which had twice the displacement of the Alfa. With more pep and far less drag because of its smaller dimensions, the Soviet model could sustain

ALFA

The smooth blending of a small sail into the hull helps make the Soviet Alfa class the fastest submarine ever built, but at high speed, the vessel is exceedingly noisy. Built of costly titanium, the Alfa has a maximum operating depth far beyond that of any adversary.

speeds in excess of forty knots, enough to outrun any rival sub—and even most torpedoes. With her titanium hull, she could descend below 2,000 feet—and perhaps approach 3,000 feet. Tactically, she appeared to be a brilliant if belated fulfillment of the late Premier Khrushchev's dreams—a machine to bury the competition.

There was just one hitch. The Alfa proved extremely noisy at high speeds—so much so that when she entered the Bering Sea on her maiden voyage, the sounds of her passing traveled thousands of miles under the polar icecap and were picked up by Navy specialists monitoring supersensitive hydrophones in Newport News, Virginia. The Alfa announced her approach so plainly at high speed that defenders could easily chart her progress on such high-priority missions as carrier pursuit. At low to medium speeds, the Alfa was reasonably quiet, but not in a class by herself.

That the designers of the world's most sophisticated sub could have ignored so serious a liability reflected Moscow's scattershot

VICTOR III

The pod atop the tail fin of this Victor III is thought to house a towed sonar array. Introduced in 1978, the Victor IIIs were the first Soviet submarines to rival U.S. contemporaries in speed, depth, and stealthiness. Ample buoyancy causes a Victor to ride high in the water, exposing the blades of its two counter-rotating propellers.

approach to sub development. Given a mandate from the Kremlin to produce a vessel that would run prodigiously fast and deep, the Alfa team concentrated single-mindedly on those goals even as advances in sonar technology and the need to seek out ballistic-missile submarines without spooking them made stealth increasingly important. The Soviets were not ignorant of those tactical trends; designers working on other Soviet attack subs had begun seeking ways to quiet them in the 1960s. Yet officials evidently shied away from a midstream overhaul of the Alfa that might have made her a less spectacular but more serviceable performer. In the end, only six of the speedsters were launched before production stopped in 1983—a small return on a huge investment.

Kremlin officials could take some solace from the relative success of the quieter designs. Built specifically for antisubmarine warfare, this Victor class—incorporating a double steel hull like the November subs but with a more streamlined exterior—went through three incarnations. The Victor I, the first of which was commissioned in 1968, sported a single propeller with an odd number of blades to minimize propeller noise. A few years later, a second version of the Victor appeared—considerably larger and, at a top speed of thirty knots, slightly slower—but distinctly quieter, thanks to a sheath of rubberized, sound-absorbent tiles that trapped much internal noise within the hull and could also deaden the echo of a ping from active sonar.

Then, in 1978, the Victor III emerged. The stealthiest of the lot—equal to the earlier Sturgeon class if not yet a match for the Los Angeles subs—this design exploited some old concepts and novel ones too. The venerable idea of mounting the sub's power plant on a noise-isolating raft was no secret by the late seventies, and intelligence analysts had evidence that the Soviets had at last begun to take advantage of this stratagem. Aerial photographs of Victors and other Soviet submarines taken during the 1980s, however, revealed something new: an intricate network of small-diameter piping installed on the hull. Those capillaries were seemingly counterproductive, since they would increase form drag, but that effect could be more than offset if the pipes somehow served to reduce skin drag.

One possibility was that they secreted lubricants known as polymers to give the sub the slipperiness of a fish. Another theory: The conduits might discharge a shroud of microfine bubbles around the hull that would prevent water molecules from touching the surface. Either approach would also quiet the sub, inasmuch as the friction of water against the metal skin produces noise as well as drag. Bubbles of air might also muffle the submarine's internal noises, since air conducts sound waves far less efficiently than water.

Precisely how hull piping might help the Soviets build swifter, quieter submersibles remains a matter of debate, but authorities agree that the successors to the Victor IIIs that emerged in the

The USS *Seawolf* was designed to be an even more potent predator than the Los Angeles-class submarines. Side-mounted sonar arrays appear as bulges on the flanks of the boat in this illustration, but they fit flush with the hull in the final design. More sensitive than towed arrays to low-frequency sounds that travel great distances underwater, the new sonar gear will dramatically increase the range at which the *Seawolf* can detect other submarines.

mid-1980s—the Akula and a similarly stealthy model called the Sierra—went a long way toward closing the acoustic gap between the United States and the USSR. Both types had double-hull construction and boasted top speeds comparable to that of the Los Angeles class; and the Sierra, with a titanium inner hull, could descend to around 2,000 feet, a depth surpassed only by the Alfas. Compounding the menace of the quiet new hunters was an improved Soviet antiship torpedo—the most lethal in history—which bore a massive 2,000-pound warhead and swam up the wake of its target at a speed of fifty knots.

However disturbing for the Pentagon, these developments in themselves did not put the Soviet fleet on a par with the U.S. fleet tactically. Granted, the Soviets, with large numbers of older diesel-electric models, had more submarines. But design problems and production delays stemming from the pursuit of unproven technologies had left them with a relatively small number of first-class attack subs. Compared to the nearly fifty Los Angeles types that the Americans could field by the late 1980s, the Soviets could muster only the six Alfas, five Akulas, and three Sierras. Furthermore, the United States maintained a wide lead in computerized passive sonar technology, meaning that Soviet subs would have to be markedly quieter than their counterparts to even the score.

Nonetheless, American skippers realized that the odds were no longer stacked as heavily in their favor. Around the time that the Akula made its quiet debut, the captain of an unidentified U.S. Navy sub, charged with tracking a Soviet boat in its home waters, got a rude reminder of how times had changed. Confident of quickly locating his quarry, he trained his ship's powerful sonar array on the surrounding sea and waited. But instead of picking up the telltale signature of a Russian sub, the Americans heard the unmistakable ping of active sonar gauging their range and bearing. The Soviets had found the U.S. sub first and fixed its location using a single sonar pulse. Had it been war, the enemy commander would assuredly have followed up his acoustic probe with a torpedo.

Anticipating that Soviet submarines would someday approach American boats in stealth, Admiral Rickover had proposed a successor to the Los Angeles class shortly before his reluctant retirement in 1982, at the age of eighty-one. Dubbed the *Seawolf* like the experimental liquid-sodium-cooled vessel of the 1950s, the one boat now slated for completion may turn out to be the only member of its class because of budget constraints. Like every American attack sub since the *Nautilus,* the *Seawolf* will carry a pressurized-water reactor, but this one will be the mightiest yet, driving two steam turbines linked to the shaft of a single propeller larger still than the one on the *Los Angeles.* Despite a displacement exceeding 9,000 tons, the *Seawolf* is expected to surpass thirty-five knots. Part of this gain will come from a close-fitting shroud around the propeller, a device that increases thrust by funneling water into the blades and by shaping the outflow into a tight stream. Noisy turbulence is reduced in the bargain.

Like stealth aircraft with their radar-absorbent paint, the *Seawolf* will be coated with a substance that soaks up sound to defeat active sonar and damps radiated noise from the hull. Additional attention to soundproofing should render her virtually imperceptible to passive sonar at speeds up to twenty knots. Even at higher speeds, her signature will be hard to detect against a backdrop of cracking sea ice, sloshing waves, or thumping drill rigs. Her capacity to elude pursuit will be further enhanced by a stronger steel hull, allowing her to dive to 2,000 feet.

Such marvels do not come cheaply. The bill for a single Seawolf submarine may run as high as two billion dollars—more than twice the going rate for Los Angeles-class subs. Critics question the need for such an exorbitantly priced supersub in an era of nuclear detente and defense cutbacks. Proponents counter that Russia, despite the demise of the Soviet Union, continues to deploy several types of nuclear submarines, including a small number of hunter-killers that can compare with anything in the U.S. fleet.

A Bitter Harvest of Betrayal

To many observers familiar with submarine technology, Soviet gains in the field during the 1960s and 1970s seemed to come so quickly that Navy chiefs and intelligence analysts wondered if breaches in security had helped erode America's technological edge. Some Soviet techniques were clearly homegrown; the United States, for example, had long since abandoned the idea of titanium hulls as too costly for the added depth capability and ruggedness

A federal agent escorts accused traitor John Walker to a court hearing in Baltimore on October 28, 1985. Both Walker and his son, Michael, pleaded guilty that morning to charges of espionage. Referring to the intelligence bonanza provided by the Walker spy ring over a period of more than fifteen years, one Soviet official crowed, "It was the greatest case in KGB history. We deciphered millions of your messages. If there had been a war, we would have won it."

they conferred and had conducted little research on the effects of polymers and bubbles. On the other hand, few believed that Moscow could have made such massive investments in silencing or reaped such dividends without some technical prompting from the West. As investigators learned, vital secrets had indeed been betrayed to the Soviets. And the damage went beyond questions of design to the core of day-to-day submarine operations—the top-secret encryption systems by which subs on patrol received orders, and whose compromise could render the fastest and quietest models critically vulnerable in wartime.

On May 19, 1985, FBI agents concealed in the roadside brush watched as Navy veteran and suspected spy John Walker got out of his Chevrolet Astro van on an isolated country road near Poolesville, Maryland, a suburb of Washington, D.C. Walker lingered a moment by a tree posted with a NO HUNTING sign, set on the ground a brown grocery bag that appeared to contain trash, and drove away. An hour later, a man identified as Soviet embassy official Gavilovich Tkachenko drove slowly past the tree and continued down the road. Suspecting that Tkachenko was casing the drop spot, the agents moved in. Packed inside the grocery bag were 129 sensitive documents detailing the operations of the nuclear aircraft carrier *Nimitz*, as well as letters and correspondence incriminating three code-named accomplices.

After the arrest of the forty-seven-year-old Walker that night, the FBI quickly set about picking up his associates. The coded evidence of accomplices merely confirmed what the agents already suspected. Earlier that spring, John Walker's estranged ex-wife, Barbara, had decided to blow the whistle on him after harboring his secret uneasily for more than ten years. At the same time, she implicated John's brother, Arthur—a former instructor of antisubmarine warfare at the Atlantic Fleet Tactical School in Norfolk,

Virginia—and another man, whom she called Jerry Wentworth. His real name was Jerry Alfred Whitworth, and he was a retired naval radioman and close friend to John Walker. Unknown to Barbara Walker, the third person implicated by the coded message—and the supplier of the 129 documents that were found in the grocery bag—was her son, Michael, an operations clerk aboard the *Nimitz.*

The collapse of the Walker-Whitworth spy ring closed a conduit that for twenty years had fed the Soviets a steady stream of secrets concerning American naval technology, operations, and communications, information that had seriously compromised both the security of America's existing underwater fleet and the Navy's plans for the future. According to Soviet defector and former KGB operative Vitaly Yurchenko, the Kremlin's espionage arm considered the spy ring the greatest espionage coup in its history—more important than the Soviet theft of Western blueprints for the atom bomb in the 1940s.

Petty desires for money and prestige had prompted this monumental betrayal. Navy men who served with ringleader John Walker remembered him as one with a taste for the highlife. While still in his late twenties, earning a modest salary as a communications watch officer, Walker had impressed his fellow seamen by purchasing a twenty-seven-foot sloop. In later years, Soviet payoffs would buy him a houseboat, a single-engine plane, a house, and a condominium. Lured by his conspicuous wealth, his brother, his son, and his best friend would all join him on the Soviet dole.

John Walker began spying in late 1967 while he was serving at operations headquarters of the U.S. Atlantic Fleet, where he was responsible for communicating with every U.S. submarine in that vast theater. Each encounter with a Soviet sub was immediately reported to his office. And if the commander of the fleet decided to redeploy his vessels against a sudden threat, Walker relayed the orders. Much of what Walker learned he passed on to the KGB. But he did even greater damage by enabling the Soviets to read Navy messages on their own.

Accorded top-secret clearance, Walker had access to the codes used for sensitive communications. Then as now, U.S. subs routinely maintained radio silence, but they had to receive messages regularly if they were to carry out their mission in a crisis. At the time, any sub awaiting orders had to deploy an antenna at or near the surface, since frequencies of existing radio transmitters could penetrate water to a depth of a few inches at most. (Subsequently, extremely low frequency radio transmitters were developed that could be used to broadcast short messages to subs at a depth of at least 400 feet, and perhaps much more.)

This constant need to communicate posed two problems—the isolated risk that a sub might be spotted near the surface as it listened for messages, and the insidious threat that a code might be broken and orders to all vessels intercepted and deciphered. To guard against such disclosures, the instructions followed by sophisticated electronic devices to encrypt and decrypt messages to subs were changed daily; without knowledge of these closely held instructions, called keys, enemy intelligence services operating even the most powerful computer would have virtually no chance of cracking a code before the coding devices were switched to another key. Walker furnished the Soviets with lists of these keys, published monthly and distributed by courier. Thus equipped, KGB computers deciphered one Navy message after another. With Walker's help, the Soviets were able to decode roughly 60 percent of all intercepted communications, gleaning a bonanza of details on U.S. submarine deployment and capabilities.

Walker's retirement from the Navy in 1976 hardly stanched the flow of secrets. By that time, he had suborned fellow radioman Whitworth, inaugurating a spy ring to which he would add brother Arthur in 1980 and son Michael in 1983. Aside from the Navy secrets his accomplices extracted, John Walker mined glittering nuggets of American submarine technology for the Soviets from unsuspecting defense contractors by posing as an industrial security consultant and planting bugs in their offices. All four members of the ring eventually received stiff prison terms.

The Navy convened a thirty-five-member committee to assess the long-term damage that Walker had inflicted. They concluded that Walker and company had been responsible for the gravest breach of national security in recent history. By enabling the Soviets to decode top-secret transmissions, Walker negated much of the gain the United States had made in stealth over the decades. Once intercepted messages had told them where to look, for example, Soviet antisubmarine warfare forces were in a position to pinpoint quiet U.S. subs that they might otherwise have missed and to develop acoustic profiles that made those vessels easier to find and track even without knowing their locations in advance.

Beyond that, the messages the Soviets were able to intercept gave them an advanced course in U.S. submarine capabilities, shortcomings, tactics, and movements the likes of which only the Navy's own submarine school could have provided. Together, the millions of top-secret communications painted a picture of U.S. underwater dominance in the submarine arena that galvanized the Soviets into action. With such persuasive intelligence to back them up, men like Admiral Gorshkov spurred the Kremlin to budget the tremendous resources necessary to develop stealthier designs at an unprecedented pace. The Victor IIIs and the Akulas were among the products of that effort.

Not all the help the Soviets received from abroad in their pursuit of stealthier subs came from spies. As Navy investigators would learn in 1986, the Soviets scored a major coup by purchasing vital technology from two firms based in countries friendly to the United States—Japan's Toshiba Machine Company and the Norwegian

A technician inspects a submarine propeller, the product of computer-controlled precision lathes. The equipment—used to fabricate nearly flawless propellers with grooved blades that reduce cavitation and its accompanying noise—is much like the machinery sold illicitly to the Soviet Union in the early 1980s by Japanese and Norwegian companies.

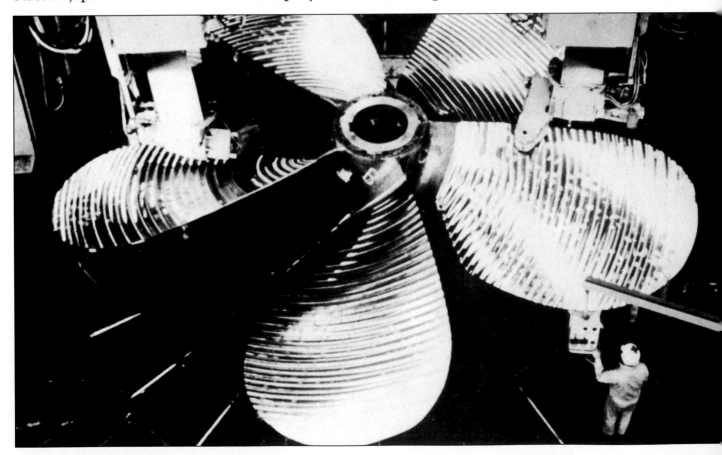

government-owned Kongsberg Vaapenfabrikk, a manufacturer of weapons and computers. Contravening an international agreement aimed at restricting the flow of high-tech devices to the Soviet bloc, Toshiba Machine and Kongsberg teamed up to sell the Soviets four room-sized precision milling machines with computer-guided cutting heads of the kind used to manufacture supersmooth propellers for American subs. To help the Soviets master the skills necessary to operate the sophisticated equipment, the manufacturers dispatched teams of technicians and software experts to the Leningrad shipyard, where they assembled the devices and provided training. According to U.S. intelligence, the Soviets used the milling machines not only to cut flawless, ultraquiet propellers for their new subs but also to make replacement props for older vessels.

Cleaning up after the Walker scandal cost the United States dearly as the Navy rushed new communications gear into service, tightened security, and pushed ahead with new designs in hopes of regaining some of the lost ground. Yet the ongoing competition took its toll on Moscow as well. Nothing the KGB or Soviet Navy gleaned from foreign sources made first-class attack subs appreciably cheaper, and as production and safety problems continued to plague the Soviet program, doubts arose in the Kremlin concerning the nation's ability to sustain an endless arms race with the West.

Even as economic and political pressures threatened to break the Soviet Union apart in the early 1990s, however, submarines were being launched from shipyards there. Assuming that the various Soviet republics emerge from the present crisis as individual members of a loose commonwealth with Russia at the fore, Moscow will likely remain the nerve center of a major nuclear power intent on guarding its shores with the keenest and quietest underwater sentries it can muster.

Propulsion without Props

For those tasked with guiding the U.S. submarine fleet into the twenty-first century, the crisis besetting America's longtime Cold War rival is no cause for complacency. While acknowledging that debt burdens or diplomatic breakthroughs could deflect Moscow from costly new development or construction programs, Navy planners must nonetheless strive for faster, stealthier designs on the assumption that idleness will inevitably dissipate America's lead in underwater technology.

Having rigorously addressed the three main sources of submarine noise—plant vibration, propeller motion, and hull flow—engineers will be hard-pressed to produce a much quieter boat with the same basic components. One way to increase stealth would be to exploit a means of locomotion—known as magnetohydrodynamic (MHD) propulsion—that does away with propellers altogether.

Scientists have long known that a strong magnetic field can make an electrolyte—a liquid capable of conducting electricity—form waves not unlike sound waves, though much more forceful and at inaudible lower frequencies. This principle has been proposed for an experimental underwater propulsion device consisting in essence of an electromagnetic coil around a tube having a rigid outer wall and an elastic inner wall, the space between them filled with an electrolyte. The hollow of the tube would be open to the sea. As a wave crest advances through the electrolyte, it would cause around the tube's inner wall a bulge that would travel the length of the tube, drawing seawater in at the front of the tube and expelling it with great force from the rear. Bulge after bulge would produce a pulsating jet of water to propel the submarine forward.

The U.S. Navy has shown little interest in MHD, in part because propelling an operational submarine would require a prohibitive expenditure of electricity in the engine's magnetic coils. However, progress in the field of superconductors—materials that can conduct electricity without loss to resistance—may eventually yield MHD engines that can propel a sub at slow to moderate speeds for missions requiring the utmost in stealth. The only sound they would generate would be the murmur of water pulsing through the intake and out the nozzle—less noise than the scything of the smoothest propellers. Although most experts believe that MHD propulsion will remain impractical beyond the turn of the century, a few dissenters disagree. They challenge the widely held view that tubular pods mounted at the stern of Soviet Akula- and Sierra-class subs house towed-array sonar—hydrophones on a cable paid out hundreds of yards behind a sub to distance the sensitive listening devices from the boat's own noise—and speculate that the pods are in reality auxiliary MHD thrusters, used in place of the propeller when silence is of the essence.

If submarines grow much quieter than they already are, passive sonar may be of little help in finding them. Some gains in sensor acuity may yet be possible using fiber optics. By replacing the metal wires that now connect towed hydrophones to the sub with glass fibers, the Navy will be able to transmit acoustic data in the form of laser light to shipboard monitors with negligible loss of volume, likely extending the range at which today's quiet subs can be detected. But should MHD or some other superstealthy propulsion technique take hold, an attack sub will stand little chance of finding a boomer before it unleashes its devastating payload. "They will be quiet, and we will be quiet," predicts Admiral Kinnaird McKee, director of the Naval Nuclear Propulsion Program. "Submarines are going to have a hell of a time finding other submarines except in the act of perpetrating some sort of violence."

Anticipating such difficulties, the Navy has explored various nonacoustic methods of submarine detection. One technique already in use—called magnetic anomaly detection, or MAD—uses magnetometers aboard fixed-wing aircraft or helicopters to pick up the faint disturbance in the earth's magnetic field caused by the presence of steel-hulled submarines in the ocean. Unfortunately, MAD does not work well against titanium-hulled Soviet subs, which have very slight magnetic properties. Moreover, when the MAD-carrying aircraft is flying high, the technique is effective only to a very shallow depth, and when the aircraft is low, it is effective over a very small area. Either way, the territory an aircraft can survey in a single flight is extremely limited, so the technique is generally used only to pinpoint submarines that have already been spotted by another method.

In the near future, space-based infrared sensors may peruse ocean surfaces for the thermal scar left in cold waters by a sub's hot reactor system. Critics of this approach suggest that its usefulness will be severely limited by the ocean's capacity to erase thermal trails as their heat is absorbed by surrounding cool waters. Certainly, it will be all but impossible to detect deep-running subs by this method.

Another proposal for nonacoustic sub detection exploits the unique capacity of light in the blue-green range of the visible spectrum to penetrate the otherwise opaque ocean. According to one scenario, a blue-green laser beamed from a satellite would sweep the ocean in search of submarines. Most of the time, the light energy would be absorbed by the ocean, but if it encountered the reflective surface of a sub, it would bounce back to the satellite.

There is some doubt, however, whether the system can be made to distinguish between subs and other large occupants of the deep such as whales. And since the diffusing effect of seawater will weaken the beam by a factor of 200 en route to a depth of 150 feet, analysts question whether lasers will be able to reach even a shallow-running sub, much less make the return trip to the satellite.

Given the dim prospects for remote detection methods, the Navy is pursuing ways of bringing passive and active sonar close to elusive targets while reducing risk to the submarine. Unmanned underwater vehicles (UUVs) may be the answer. Stowed in the sub's hull, quiet, compact UUVs would be guided by means of long fiber-optic links to points miles away from the mother sub. There they could serve as acoustic pickets or as decoys broadcasting a dummy acoustic signature to draw fire from enemy subs, which would betray their positions by the sound of torpedoes being shot from their tubes. Conceivably, autonomous UUV decoys could be programmed to range much farther afield. From scores—or even hundreds—of miles away, the watchful mother sub, using passive sonar, could hear the enemy firing at the UUV and either pass on the target's location to friendly sub hunting forces in the area or maneuver in quietly for a torpedo attack.

How many of these options will be actively pursued—and how soon they will be operational—depends in part on the future of East-West relations. In all likelihood, Moscow will continue the trend of recent years by significantly reducing the size, if not the capability, of its submarine force. The U.S. Navy, for its part, may see its own fleet of attack submarines shrink as more boats are retired than are built to replace them. One thing seems certain: So long as nuclear submarines exist, rival crews assigned to prepare for the unthinkable will shadow one another in the abyss, straining to make out the foe's faint signature amid the gloom of the sea. ★

Whispers in the Sea

The moment a submarine slips below the surface, sensory impoverishment sets in. Because the sea is largely opaque to light, radio waves, and other forms of electromagnetic radiation, virtually all knowledge of what lies outside a sub must be gleaned from sound, the one form of information-bearing energy that travels easily through water. Sound consists of molecular vibrations that expand outward like concentric ripples. Since molecules are much more closely packed in water than in air, sound moves through the sea about five times faster—an average of 3,350 miles per hour.

Undersea detection by sound—called sonar, for sound navigation and ranging—takes two forms, active and passive. Chiefly exploited by surface ships in search of submarines, the active mode emits sound as high-intensity pings that reflect off an object and reveal its whereabouts. Subs most often use sound passively, monitoring the acoustical environment with ultrasensitive detectors and powerful computers.

Though seemingly simple, stalking by sound is fraught with difficulties. Background noise is generated by ships' propellers, waves, volcanic activity, shrimps' claws, whale calls, and the like. Moreover, sounds are scattered by clouds of microscopic creatures and are reflected off the sea's surface, its floor, and boundaries of regions with sharp differences in temperature. Except under polar icecaps, where water temperature is constant regardless of depth, the sea has a layered architecture that determines the speed of sound. As explained on these pages, sound velocity varies with water temperature and pressure. Cooler temperatures lower speed while higher pressures increase it. These influences refract, or bend, sound and, in deep water, force it to travel in curved paths that can mask an enemy presence utterly—or reveal it at great distances.

The sea's surface layer, though fairly uniform in temperature throughout its depth (as much as 200 feet), gains and loses heat in response to influences such as warmth from the sun and changes in season.

In the next layer down, called the main thermocline and extending to a depth of about 2,000 feet, the temperature of the water drops steadily, reducing sound velocity more than rising pressure raises it.

At the bottom of the main thermocline, where the water temperature stabilizes at about 39 degrees F., the velocity of sound reaches a minimum.

Extending from the main thermocline to the seafloor is the isothermal layer. In this region of constant temperature—also called the deep sound channel—sound velocity rises as pressure increases with depth.

SURFACE LAYER

MAIN THERMOCLINE

POINT OF MINIMUM VELOCITY

ISOTHERMAL LAYER

Curving Paths of Propagation

Just as a prism bends light by slowing it down, sound is bent by change in its velocity, refracting toward the region of slower travel. Of the two chief determinants of speed, temperature has the greater effect, except in the uniformly cold water below icecaps. For each degree temperature drops, speed falls by about 5 feet per second. Greater pressure increases velocity, but at a rate much smaller than the decrease from lower temperatures. Thus the higher pressure 100 feet deeper in the thermocline—a distance that drops temperature about one degree—raises speed by only 1.8 feet per second.

Any noise from a submarine propagates outward in all directions, weakening as it travels. Refraction, reflection from the surface and boundaries between ocean layers, and scattering caused by plankton complicate its path, but key patterns of propagation can be delineated by tracing a few sound rays *(right)*. A remarkable acoustical phenomenon is a sinuous, long-range form of travel caused by a repeating cycle that begins with falling temperature bending sound downward and continues with rising pressure curving it back upward. Sound propagated in this way—called a convergence zone (CZ) path—rises periodically until all its energy has been lost to the sea.

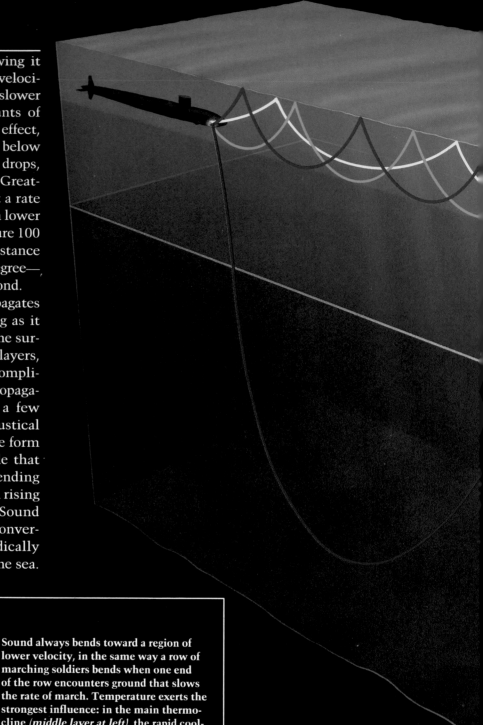

Sound always bends toward a region of lower velocity, in the same way a row of marching soldiers bends when one end of the row encounters ground that slows the rate of march. Temperature exerts the strongest influence: in the main thermocline *(middle layer at left)*, the rapid cooling with depth causes velocity to fall and sound to be refracted downward. Eventually—at the point of minimum velocity—temperature ceases to drop and pressure takes control. The steadily increasing pressure in deep sound channel *(bottom layer)* raises the velocity again and gradually bends the sound upward.

Four rays of noise from a submarine in the surface layer are traced below. Because temperature is uniform in this layer, the increase of pressure with depth refracts the blue, green, and yellow rays upward; they then reflect downward from the surface and are bent upward again repeatedly, as though traveling in a duct. The steeply falling red ray, which represents a band of rays traveling even more vertically, plunges deeper, where opposing refraction by temperature and pressure produces a snakelike CZ path rising to the surface every thirty miles or so.

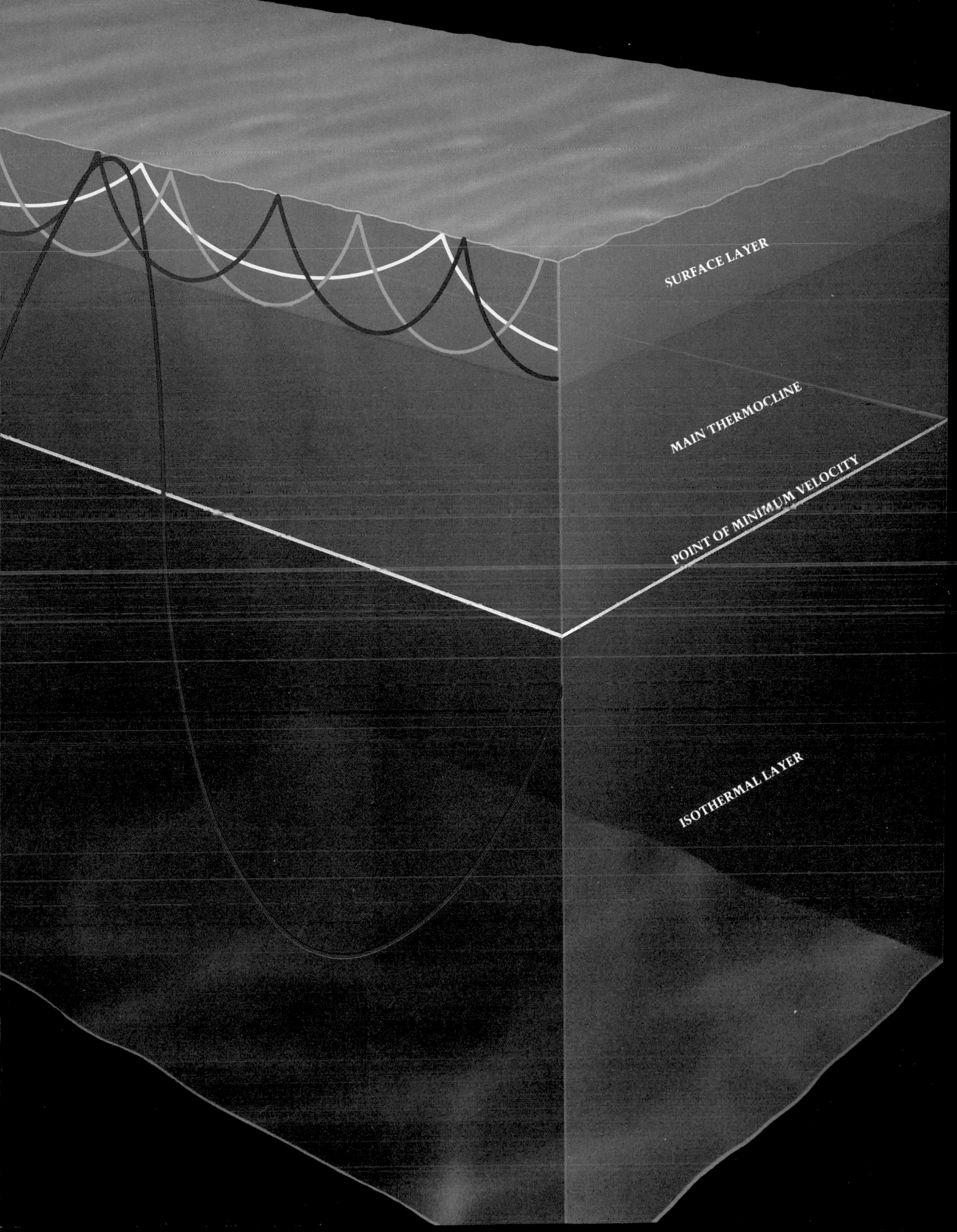

SURFACE LAYER

MAIN THERMOCLINE

POINT OF MINIMUM VELOCITY

ISOTHERMAL LAYER

69

The Importance of Position

Whether submarines can hear one another depends on their relative positions in the vertical layering of the sea as well as on the horizontal distance separating them. If a sub is in the surface layer, much of its own noise will be confined there by reflection and upward refraction, contributing to the loudness of the environment and obscuring the telltale sounds of any opponent. Some of the sound, traveling at a steep angle, will penetrate downward, where it may be detected by a lurking foe. Meanwhile, most of the radiated noise of a deeper submarine—or one at any depth under a polar icecap—is refracted downward, producing a quieter listening environment and helping to conceal the sub from an enemy above.

But propagation paths are never simple. Because some noise will be scattered by particles, bubbles, and other obstacles, no region is perfectly safe. Regardless of vertical placement, contests of hide-and-seek are generally won by silence and keen hearing.

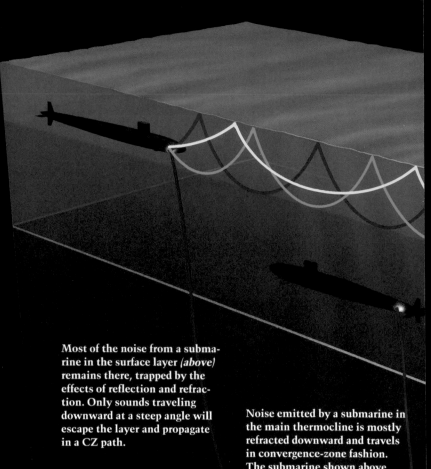

Most of the noise from a submarine in the surface layer *(above)* remains there, trapped by the effects of reflection and refraction. Only sounds traveling downward at a steep angle will escape the layer and propagate in a CZ path.

Noise emitted by a submarine in the main thermocline is mostly refracted downward and travels in convergence-zone fashion. The submarine shown above cannot detect the sub in the surface layer—and vice versa—because their positions do not coincide with the CZ path.

One sub near another can hear murmurs from its opponent by direct propagation, rather than by the convergence-zone path shown in the large illustration at right. Possible between two subs in the main thermocline, as shown here, as well as in the surface layer, this type of detection generally has a range no greater than about five miles. In such settings, stealth and the acuity of sonar gear govern which detects the other first, since acoustical conditions are equal.

The illustration below shows how a submarine in the main thermocline *(far right in the drawing)* can hear opposing submarines in some locations but not in others. Detection range is typically much greater when a sub is eavesdropping on sounds traveling by CZ path than on those traveling by direct propagation. Moreover, since sound can travel both ways on a CZ path, noise coming from an eavesdropping submarine can reveal its presence to the opponent.

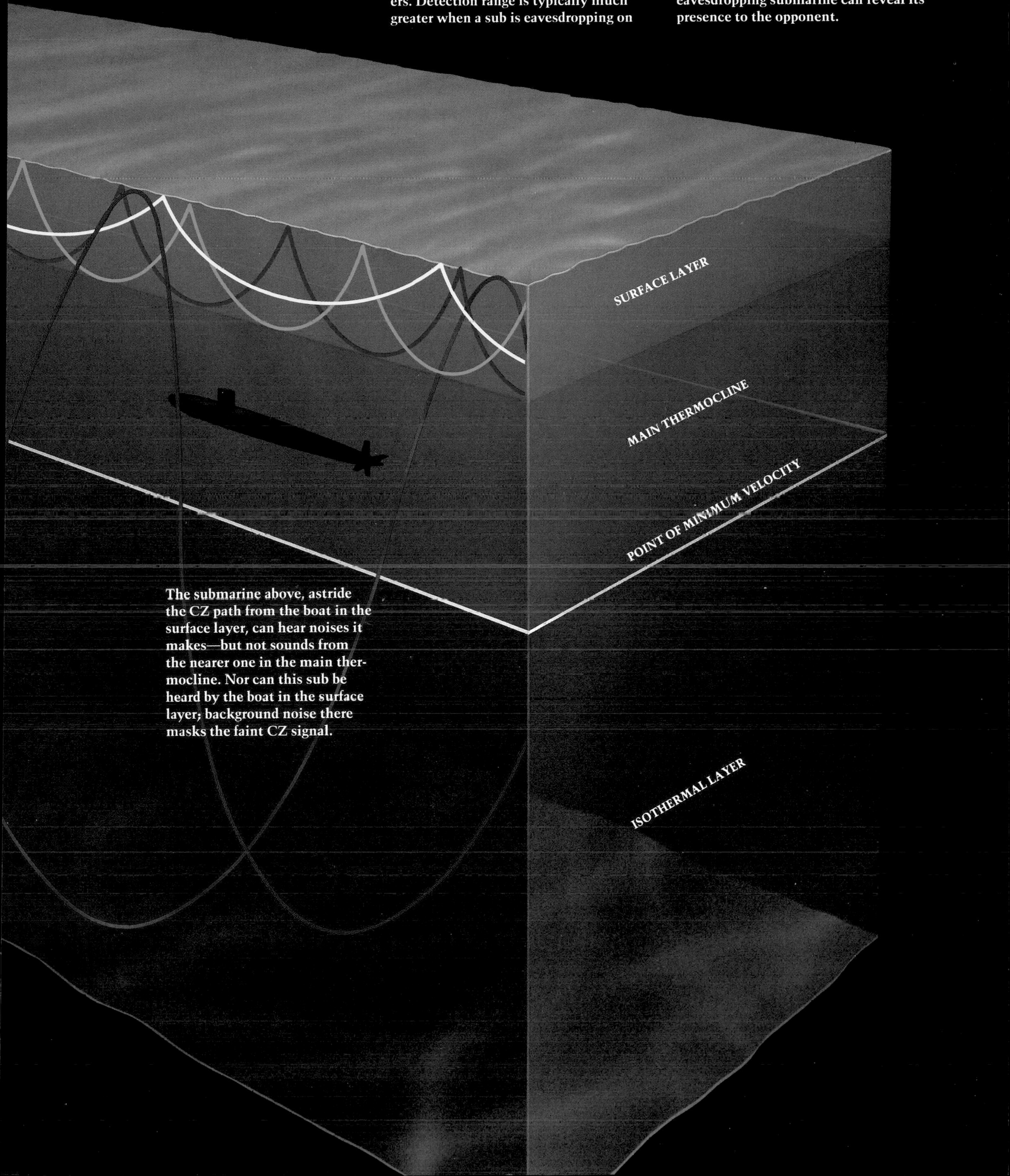

SURFACE LAYER

MAIN THERMOCLINE

POINT OF MINIMUM VELOCITY

The submarine above, astride the CZ path from the boat in the surface layer, can hear noises it makes—but not sounds from the nearer one in the main thermocline. Nor can this sub be heard by the boat in the surface layer; background noise there masks the faint CZ signal.

ISOTHERMAL LAYER

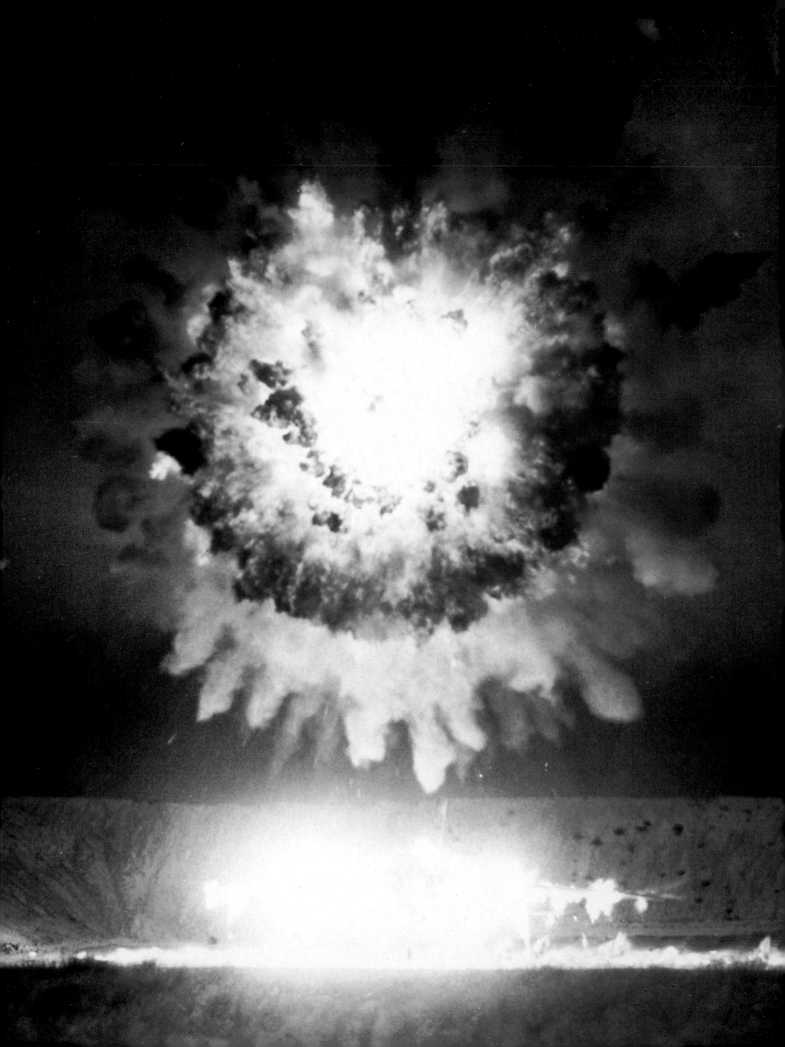

3

The Long Reach of the Submarine

Launched by a submerged Los Angeles-class submarine 400 miles away, a Tomahawk cruise missile detonates above an aircraft during a 1986 test at San Clemente Island in California. The explosion scattered bomblets by the score, some of which struck the plane and set it afire.

Long, low, and sinister-looking in her matte black paint, the nuclear radar-picket submarine USS *Triton*—SSRN 586—slipped out of the sub base at Groton, Connecticut, and eased down the Thames River on her way to sea. In the ship's log, the date read February 16, 1960. The boat had just been commissioned as the U.S. Navy's most advanced nuclear-powered sub, and her crew was about to embark on what they expected to be a routine shakedown cruise. They learned differently when the ship reached the open Atlantic. Captain Edward Beach gave the command to submerge and then announced the Navy's stunning order for this mission. The *Triton*, Beach declared, had ventured forth on "the voyage which all submariners have dreamed of. We are going to go around the world, nonstop. And we're going to do it entirely submerged!" If successful, the *Triton*'s voyage would demonstrate a quantum leap in submarine capabilities: From now on, submarines could patrol and fight for months at a time anywhere in the world without ever having dry decks.

Nuclear ship propulsion, though still a dawning field in 1960, made the *Triton*'s reenactment of Ferdinand Magellan's sixteenth-century circumnavigation of the globe a feasible goal. Diesel-powered submarines could remain submerged for long periods of time by means of a snorkel—a twin-tubed device protruding just far enough above the surface to provide fresh air and an exhaust outlet for the engines. Even so, diesel subs could range only as far as their fuel tanks would permit—and none could sail around the world without surfacing a number of times to refuel from an oiler.

The 5,940-ton *Triton* would serve as an excellent vehicle for the experiment. Designed to roam ahead of the fleet at periscope depth with a radar antenna extended into the air, she boasted twin water-cooled nuclear reactors in her roomy 450-foot hull. She had space both for additional crewmen and for equipment to be tested. Scientists from the Navy Hydrographic Office would be on board to conduct numerous oceanographic tests. The *Triton*'s voyage would be a trial run for several important pieces of equipment that naval engineers hoped to install in future nukes. Captain Beach could obtain a celestial navigation fix by shooting the stars through a periscope of new design, obviating the need to surface and employ a sextant. The *Triton* also mounted the first production model of a ship's inertial navigation system (SINS), which was expected to keep track of a sub's position automatically and with unprecedented precision. Yet another experimental device on board, the precision depth recorder (PDR), was coupled to the boat's standard Fathometer. The PDR graphically traced depth soundings on special, sensitized paper to produce a profile of the bottom.

More than just an evaluation of hardware, the *Triton*'s cruise would serve as a study in human behavior. No sub crew had ever been confined for so long a time without putting in to port; but this sort of extended, isolated duty would become the norm for nuclear submariners in the not-too-distant future. A Navy psychologist accompanied the *Triton* on her voyage to observe and evaluate the effects of the mission on the crew's mental health. Captain Beach set up instructional and entertainment programs for the men in their off-duty hours, and initiated a ship's newspaper to mitigate the absence of mail from home and lengthy isolation from the world.

The prototype PDR passed its first test impressively, mapping previously uncharted underwater mountains, as the *Triton* ran southeast toward St. Peter and St. Paul rocks, a pair of isolated outcroppings in the equatorial Atlantic off Brazil. This landmark would be the formal starting point for the round-the-world venture.

On February 24, Captain Beach circled the barren islet, giving everyone a turn at the periscope, then shaped a southwesterly course down the South American coast. Twelve days later, the *Triton* rounded Cape Horn and headed northwest into the vast Pacific on a track that closely paralleled Magellan's historic voyage more than 400 years earlier. Various operational exercises punctuated the endurance cruise; the *Triton* slipped into U.S.-patrolled waters of the Philippines and Guam to photograph harbors and naval bases as a reconnaissance exercise and to test her nuclear-age stealth capabilities against antisubmarine warfare (ASW) forces.

Not once did American sonar detect the submarine. As the *Triton* sailed around the Philippines, however, a fisherman spotted the sub's periscope rising from the depths. Beach learned later that the superstitious man thought he had stared a sea monster in the eye.

On sailed the *Triton*, across the Indian Ocean, around Africa's Cape of Good Hope, and finally, on April 25, back to St. Peter and St. Paul rocks. A quick dash across the North Atlantic brought the thoroughly tested nuclear sub back to home waters. Except for the PDR, which stopped working after the Fathometer failed about halfway through the voyage, all systems had passed their trials, and the crew had performed flawlessly. The captain's log entry for May 10 reads: "Surfaced, having been submerged exactly 83 days and 10 hours, and traveled 36,014 miles. The men of the *Triton* believe her long, undersea voyage has accomplished something of value for our country. The sea has given us a means of waging war, but even more, it has given us an avenue to hold the peace."

Captain Beach's words would prove prophetic. In the decades to come, the nuclear ballistic-missile submarines of the superpowers, running deep, silent, and deadly, would serve as powerful deterrents to any all-out conflict. Meanwhile, the modern attack submarine, both nuclear and diesel-electric, would find innumerable uses as the journeyman combatant in the navies of the world.

Aside from its function as the most effective hunter of its own kind, the submarine would play a dramatic role in the aggressive games of the Cold War. Submarines of the NATO and Warsaw Pact navies would engage in never-ending reconnaissance and other covert operations against their putative foes. Though no one would say much about it, both sides would stalk each other's surface ships to perfect tactics and techniques, indeed, to perform every drill short of actually shooting at one another.

As technology advanced, submarines would find employment in conflicts ranging from India to the Falkland Islands. And in the recently concluded Persian Gulf War, American nuclear attack subs—SSNs in U.S. Navy nomenclature—would for the first time employ a new cruise missile technology that would extend their reach across 800 miles of sea and land. Throughout, in every operation, with every kind of submarine, the watchword would be secrecy. "The whole name of the game is stealth," declares Captain Fred P. Gustavson, a former Sturgeon-class attack sub skipper.

"Submariners do not fight fair. If it turns into a fair fight, the submariner ought to withdraw and come back another time."

Prowling the Waters of Neutrals

Like the Filipino fisherman in 1960, the Swedish trawler captain could not believe his eyes as he motored up the Bay of Gaase on the morning of October 28, 1981. There, in a narrow channel not twenty miles from Karlskrona, one of the country's main naval bases, lay a large submarine with its bow run up on the rocks. A great racket reverberated across the water as the sub revved its diesel engines to maximum speed in a vain attempt to work itself free.

The fisherman reported the peculiar sighting to Swedish naval authorities, who at first discounted his story as that of a crank or a drunk. Not until twelve hours later did Swedish naval vessels arrive to investigate. They found the sub—a Soviet Whiskey-class diesel-electric boat—exactly where the fisherman had said it was, still furiously trying to extricate itself.

When challenged, the skipper of the Whiskey, Captain Second Rank Anatoly Gushchin, feigned wide-eyed innocence. He explained to the incredulous Swedes that he had experienced problems with his navigational and radar equipment. The breakdown, he said, had led him off course. Gushchin told the Swedes that he thought he had grounded on the coast of Poland and expressed astonishment to learn that he was deep inside Swedish territorial waters near a sensitive naval base eighty miles on the other side of the Baltic. That, of course, was ridiculous. Gushchin could never have navigated the miles of narrow, twisting channels through the islands near Karlskrona to the point where he

An American flag flying from a mast atop the sail, the USS *Triton* returns to port in Groton, Connecticut, following its record eighty-three-day around-the-world journey underwater. The sub has an unusually large sail, designed to hold radar equipment for the boat's intended role as an advance scout for the fleet.

came to grief unless all his equipment had been in perfect working order. Rather obviously, the Soviet sub had been on a reconnaissance mission when it blundered aground.

Sweden impounded the Whiskey where it lay as diplomatic notes flew back and forth, the Swedes protesting angrily, the Soviets indignantly declaring their innocence and demanding the release of the sub forthwith. At one point, the Swedish ambassador to Moscow appeared before a Soviet deputy foreign minister—to find ten glowering admirals ranked behind the minister's desk.

Eventually, the Swedes acquiesced. Tugs worked the sub free, towed it out to sea, and sent Captain Gushchin home—to serve, it is said, several years in a gulag for his bumbling. Boarding the Soviet vessel, the Swedes established from the clumsily altered log that the boat had been snooping around in the area for three days, apparently as part of a larger reconnaissance operation, and had on board a submarine squadron commander. Further, and of great concern, the Swedes discovered by means of a radiation analyzer that the boat carried as much as twenty pounds of fissionable U-238. Inasmuch as the Whiskey had no atomic power plant, the Swedes deduced that the sub carried at least one nuclear weapon.

Western newsmen naturally headlined the incident "Whiskey on the Rocks." Yet there was nothing to snicker about. The Scandinavian peninsula, sitting athwart the Soviet Navy's only access route to the North Atlantic from the Baltic Sea, occupies a position of the utmost strategic importance. Norway's long, deeply indented coastline provides plentiful anchorages for NATO ships, while the Kattegat channel between Denmark and Sweden is a natural choke point at the only entrance to the Baltic Sea and the port of St. Petersburg. It could come as no surprise that the Soviets would mount a close watch on the harbors, inlets, and military installations that dot the coastline.

Denmark and Norway, as members of NATO and potential foes in any future hostilities, could be considered fair game for this sort of activity. Sweden, on the other hand, had long stood aloof from both NATO and the USSR—and had built potent armed forces to defend this independence. Yet despite Sweden's impeccable neutrality, Soviet submarines continued to violate Swedish waters. Instead of pulling in their horns after the 1981 Karlskrona debacle, the Soviets continued their efforts.

A year after the Whiskey episode, Swedish ASW forces detected another massive Soviet recon operation, involving at least three full-sized submarines and three minisubs, one or more of which, judging from evidence on the seabed, appeared to be a tracked submersible capable of crawling along the bottom. This time, the Soviets targeted the huge naval complex on Musko Island in Haar Bay near Stockholm. ASW forces chased the subs around for most of October, dropping more than forty depth charges near the uninvited visitors, not to try to sink them but to make them surface. But the Soviets brazenly continued with the mission. Later, divers found track and keel depressions on the muddy seafloor, indicating that minisubs had penetrated to the port of Stockholm itself.

Yet another bold operation took place in early 1984, again at Karlskrona. By now, the Swedes had laid magnetic anomaly detectors on the seabed; the sensors picked up the first intruder on Feb-

In 1981, while spying on Sweden, this Soviet Whiskey-class submarine ran afoul of the rugged Swedish coastline. Heavy seas battered the vessel, tilting it seventeen degrees and threatening to puncture the hull against the rocks.

ruary 8, and before the Soviets finally departed in early November, no fewer than 600 detections had been recorded by the MAD gear and by hydrophones, sonar, and surface-radar and visual sightings of periscopes. The Swedes believed that at least three and perhaps five fleet submarines were involved, along with minisubs, tracked submersibles, and divers. As before, ASW forces dropped depth charges to no effect—and Swedish shore units responded to at least five reported landings by Soviet frogmen, three times firing warning shots at swimmers seen leaving the water on the island of Almo.

Since the Swedes never captured anyone or forced any boats to surface, the Swedish government could officially maintain that the intruders were unidentified. But in 1987, as the incidents continued to multiply, General Bendt Gustavson, commander in chief of Sweden's armed forces, abandoned that conceit. "The Soviet Union," he said bluntly, "is behind the submarine incursions." By the end of 1988, Swedish ASW-force records dating from 1962 would show 191 instances in which foreign subs, presumably all Soviet, had undeniably penetrated Swedish waters. Some 200 additional cases were judged slightly less well proved. In 1984 alone, the Swedes detected 60 incursions. Such numbers are a measure of how greatly the Soviet Navy valued its submarines for such assignments.

Lest the Pot Call the Kettle Black

Soviet subs were far from the only players in this game. American boats—referred to in one of the rare newspaper articles on this supersecret topic as "underwater U-2s"—assiduously worked the USSR's coastal waters for intelligence throughout the Cold War. The full extent of their clandestine activities may never be known, precisely because the sub is so good at not getting caught—or getting away if detected. Yet there have been some highly uncomfortable moments. In August 1957, the USS *Gudgeon*, an American diesel-powered boat on an intelligence-gathering mission, learned how it feels to be trapped. The *Gudgeon* had stationed herself in shallow water barely more than three miles outside the port of Vladivostok, site of the largest Soviet naval complex in the Pacific. Technically, the Navy could claim freedom of the seas since the United States recognizes only a three-mile limit; however, the Soviets claimed a twelve-mile limit and meant to enforce it.

The *Gudgeon* had a double objective: to note the arrival and departure of Soviet warships, and to monitor radio communications to and from the base. To this day, no one on the *Gudgeon*'s crew knows exactly how the Soviets discovered the sub, but suddenly Vladivostok's communications channels started sounding an alert. Within a few minutes, no fewer than eight destroyers came boiling out of the harbor on a beeline for the *Gudgeon*'s position.

Lieutenant Commander Norman B. Bessac sent his ninety-man crew to battle stations, flooded the eight already-loaded torpedo tubes, and turned for the open ocean. Having no more than half the speed of the destroyers, the *Gudgeon* was swiftly overhauled. In deeper water about twelve miles out to sea, Bessac took his vessel down to 200 feet and brought it to a full stop. He ordered all non-essential machinery turned off and cautioned the crew to stay quiet. By minimizing noise, he hoped to throw the pursuers off his trail.

But the Soviets had the range, and the Americans could hear the loud pings as sound pulses from the destroyers' active sonar hammered the *Gudgeon*, pinpointing her position. Then came the terrifying thunder of depth charge salvos detonating nearby. "They would circle and one of them would make a run," a sailor recalled of the destroyers' explosives. "Then they'd go back out and pick us up again and somebody else would come in and make a run."

After the first bracketing salvos failed to cause any serious damage, the *Gudgeon*'s officers concluded that the Soviets at least for the moment felt like toying with them, dropping practice depth charges with insufficient explosive force to rupture the submarine's thick steel hull. That gave little comfort to the men in the trapped sub. As one crewman put it, "The thing you worry about when they drop the damn things is that the next one's going to be a real one."

Several times, Bessac ordered the *Gudgeon* forward at the slowest speed possible, hoping to escape the circle of destroyers, but sonar pings tracked every movement, and the Soviet ships laid fresh strings of charges across the boat's path. At one point, Bessac sent out a noisemaking decoy through the sub's garbage ejection tube; the Soviet ASW captains ignored the deception. Growing desperate, Bessac took the *Gudgeon* below its 700-foot maximum operating depth; despite some creaking and groaning, the *Gudgeon*'s hull took the pressure, but Soviet sonars still found their target. After thirty hours under siege, with batteries dangerously weak and the air in the boat thoroughly foul, Bessac had few options left.

He brought the *Gudgeon* to periscope depth, quickly raised his antenna, and flashed a distress call to U.S. Seventh Fleet headquarters in Japan. Bessac also extended the snorkel, hoping to suck in enough fresh air to remain submerged. Immediately, a Soviet destroyer came racing straight for his position, apparently with the intention of ramming the snorkel tube. Bessac went deep again, but he knew that he would soon have to come up for air.

Game over, the *Gudgeon* surfaced, torpedo-tube doors open and ready to fire. Bessac climbed up into the sail and ordered a signalman to flash a message in morse code to the nearest Soviet destroyer. Bessac identified his vessel as a U.S. Navy warship and asserted his belief that the *Gudgeon* lay beyond the twelve-mile limit. Back came a curt signal advising the Americans to depart forthwith. With that—and no doubt a clear sense of having won this round—the ring of Soviet destroyers parted and allowed the *Gudgeon* to withdraw.

Bessac had the misfortune to be discovered; most other U.S. subs on similar missions are not. According to some estimates, American submariners have performed more than 2,000 intelligence-gathering sorties since the late 1940s, all but a handful of them in absolute secrecy. Although submariners are a tight-lipped lot, some have hinted that American sub skippers have photographed the maiden voyage of virtually every new Soviet surface vessel or submarine—sometimes at such close range that three images had to be joined to complete a montage of the hull. Moreover, underwater photographs can be taken through a periscope at shallow depth in daylight, and there are stories of U.S. subs passing less than fifty feet beneath Soviet ships to photograph the hulls and tape propeller sounds to establish an acoustic signature for future reference.

Some of these tales of derring-do may be apocryphal—and some perhaps not. According to former crewmen, in 1961 the USS *Harder*, a diesel-electric submarine, actually penetrated the channel leading to the naval complex at Severomorsk, near Murmansk on the Soviet Arctic coast. The *Harder*'s skipper reportedly photographed the piers, drydocks, and ways of this major shipbuilding center before silently departing. "It seemed like forever, but the run was probably less than a hour," said a member of the crew. Later, in the 1980s, a nuclear-powered American attack sub reputedly sneaked even farther up that same channel almost to the port of Murmansk itself, drifting along at "about half a knot," reported one crewman, while snapping photographs close enough to shore that faces could be made out through the boat's periscope.

On at least one occasion, however, a U.S. nuclear sub on a recon assignment has found itself in a confrontation akin to that of the diesel-powered *Gudgeon* off Vladivostok. As it turned out, the encounter spoke volumes about the capabilities of the new American SSNs. In 1963, six years after the *Gudgeon's* ordeal, the attack sub *Swordfish* approached a Soviet naval exercise near the Kamchatka Peninsula in the northwest Pacific. As the *Swordfish* photographed formations and scooped up communications intelligence with its antenna, a lookout on one of the Soviet warships spotted a glint of light from the sub's periscope.

Several ships swung toward the sighting, while the *Swordfish* went deep. As one crewman remembered, the Soviets "spent the better part of two days dropping depth charges here and there." Mostly, however, the *Swordfish* held the initiative. Unlike the *Gudgeon's* Lieutenant Commander Bessac, the *Swordfish's* skipper did not have to worry about depleting his batteries or air supply during an extended underwater engagement. Frequently during the encounter, the *Swordfish* would come to periscope depth to observe the Soviet ships and to record ship-to-ship communications, as well as radar frequencies and search patterns. In this manner, the boat collected a bonanza of intelligence data on Soviet antisubmarine warfare techniques. "We got to know exactly what the Soviets' capabilities were," a crewman confided. When they returned to Pearl Harbor, the entire crew received Navy commendations, and the captain of the submarine was awarded the Legion of Merit.

Tapping the Wire

At times, covert intelligence missions have demanded more than simple observation. In the early 1970s, the National Security Agency and the U.S. Navy initiated a venture, code-named Ivy Bells, to intercept Soviet military communications passed by undersea cable between bases on the heavily fortified Kamchatka Peninsula and

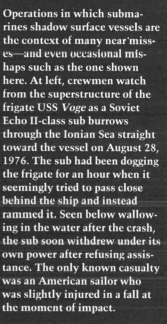

Operations in which submarines shadow surface vessels are the context of many near misses—and even occasional mishaps such as the one shown here. At left, crewmen watch from the superstructure of the frigate USS *Voge* as a Soviet Echo II-class sub burrows through the Ionian Sea straight toward the vessel on August 28, 1976. The sub had been dogging the frigate for an hour when it seemingly tried to pass close behind the ship and instead rammed it. Seen below wallowing in the water after the crash, the sub soon withdrew under its own power after refusing assistance. The only known casualty was an American sailor who was slightly injured in a fall at the moment of impact.

the Soviet Pacific Coast mainland. The USS *Halibut*, originally designed as a cruise-missile sub, was fitted with special thrusters that allowed her to remain motionless underwater. At the appointed time, she stole silently into the Sea of Okhotsk, which separates Kamchatka from the Soviet mainland, and descended 400 feet to hover above the cable lying on the seafloor. While in that position, she sent out specially equipped divers who attached an eavesdropping device to the cable. After the *Halibut* was retired in 1976, the Navy equipped two other nuclear submarines, the original *Seawolf* and the *Parche*, with thrusters and periodically sent them back to collect tape-recorded information from the undersea wiretap.

The Soviets never detected the subs and had no idea that their secure channels had been compromised until a disgruntled former

Trolling for Submarines

A U.S. Navy P-3 Orion patrol plane flies over a Soviet Victor III-class nuclear submarine, crippled and forced to surface in 1983 after an entanglement with a towed-array sonar trailed by the frigate USS *McCloy*. The incident occurred some 470 miles east of Charleston, South Carolina, where Soviet subs regularly monitor American operations. Usually they keep clear of the Navy ships; this time, however, the Victor caught its propellers in steel towing cable, which stretched hundreds of yards behind the *McCloy*. A Soviet tug dispatched from Cuba arrived a few days after the encounter to take the Victor in tow.

U.S. intelligence official revealed the information in exchange for cash. His treason scuttled one of America's most successful peacetime covert operations.

A number of years later, the Soviets apparently suffered a disaster while attempting a kindred operation. Exactly where the action took place remains classified. However, in the summer of 1985, the U.S. Navy detected a Soviet Zulu-class diesel-electric boat loitering on the surface in the North Atlantic. The Zulu was in international waters and had violated no law. Nevertheless, the Navy sent the Los Angeles-class nuclear attack sub *Baltimore* to investigate.

At periscope depth and moving very slowly, the *Baltimore* made a first pass within 100 feet of the Zulu. Raising the periscope—but not enough to break the surface—the skipper observed the twin hulls of a catamaran-style sled in the water next to the Soviet ship. The sled had ballast tanks for descending and rising. Extending below the water line was a scoop-like device, presumably for digging at the ocean floor 300 feet below. Crewed by men in deep-diving suits, with air hoses connected to the sub, the sled disappeared below the surface. Murky water prevented the skipper from counting the divers.

"They had no idea we were there," said one *Baltimore* crewman as his submarine maintained its watch from a safe distance. As a further precaution, the U.S. boat took steps to reduce its own noise. "Everybody told us to be quiet, wear your rubber shoes, don't slam doors. They turned off the ice machine—it grinds ice and makes noise. They turned off the soda machine."

Meanwhile, the *Baltimore*'s passive sonar picked up sounds from the sled. "They were making a lot of noise that was repetitive. It sounded like digging and drilling," explained the crewman. The *Baltimore*'s officers surmised that the Soviet divers hoped to locate and tap into American communications cables buried in the bottom. If so, the effort stands as a model of ineptitude in allowing itself to be so easily discovered.

For the next three days, the *Baltimore* studied the Zulu—even listening in via sonar on communications between the boat and the divers. Meanwhile, the weather deteriorated until the swells reached thirty feet in height. That posed no problem for the submerged *Baltimore,* but the Zulu plunged and rolled on the surface.

Then the noise from the sled abruptly ceased. "We were quite a ways away so we couldn't be seen," related a *Baltimore* crewman. "When we heard the Zulu start up again," he continued, "we came closer for a better look and there was no more sled." The tethering cable, most likely severed by the turbulence of the towering waves, hung limply in the water.

"I remember that everyone in the conn turned around and looked at each other," said a *Baltimore* crewman, recalling the silence in the control room. Though the Soviets remained adversaries whose mission had failed, no one on the *Baltimore* felt a sense of triumph. "It was more like we realized a submariner was dead."

Booby Traps of the Deep

Few weapons offer the navies of the world a greater force multiplier than mines. Sown across a coastal shipping lane, the entrance to an enemy harbor, or a strategic ocean choke point, just a handful of them can create havoc. Virtually any vessel and many types of aircraft can lay mines, but they cannot do so surreptitiously. As Gunners' Mate First Class Jeffrey K. Bray puts it: "To effectively deploy a clandestine minefield, the delivery vehicle must remain concealed. The ideal vehicle for this is the submarine."

Mines must be specially configured for delivery through a sub's torpedo tubes. Some smart mines are designed to destroy surface vessels, while others are specifically engineered as sub killers. The U.S. Navy's CAPTOR consists of a standard Mark 46 antisubmarine torpedo fitted into a special casing that houses acoustic sensors and a microcomputer *(pages 148-149)*. In the event of war, U.S. attack subs, each carrying a ration of CAPTORs, would seed them in areas where Soviet boats were known to transit. The submarine-launched mobile mine (SLMM), intended mainly for surface vessels, has acoustic, pressure-sensitive, and magnetic triggers. Its mo-

A few U.S. submarines have been modified specifically for special operations. Here, crewmen of the USS *John Marshall* signal the successful launch of a SEAL delivery vehicle (SDV), seen floating above them. A submersible that can transport up to six people, the SDV rides in a tubular hangar that is flooded just before launch. After opening the hangar's circular door and extending a set of tracks, SEAL divers push the SDV outside. With its complement of SEALs aboard, the SDV is then released from the track to assist in missions that include reconnaissance and attaching mines to the hulls of ships.

bility makes it an ideal sub-launched weapon. Fired like a torpedo, it swims more than five miles to its intended destination; when it arrives, it sinks to the seafloor and lies in wait for a passing ship.

An effective minefield can consist of as few as eight or ten mines. Usually, it would be wasteful to plant more, since at the first detonation the enemy obviously would reroute shipping until the area could be swept clean. However, some mines actually bury themselves in the bottom mud to avoid detection; others combat minesweeping operations with ship-counting mechanisms that allow three, four, even a dozen vessels to pass overhead before reacting. With such devices to contend with, an enemy is likely to reduce his activities on the mere suspicion that a minefield exists around a port or choke point.

"Mine warfare is a business where you can't afford even one little screw-up," says Captain Fred Gustavson. For example, mines cannot discriminate between friend and foe; once they are armed, mines become as deadly to one as to the other. Thus Gustavson, like his fellow attack-sub commanders, makes meticulously detailed plots of the mines he plants.

Although laying minefields is an important responsibility, submariners do not relish the task. For one thing, subs are never around to witness the results of their minelaying handiwork. At best, the crew may hear, far after the fact, of an enemy ship that blew up about where they had planted a mine. They would be unlikely ever to learn of an enemy sub sunk by a CAPTOR. Furthermore, mines take up precious space that could otherwise be occupied by torpedoes, and all things considered, submariners would rather shoot at an enemy craft with torpedoes than waylay it with mines.

Stalking the Heavy

In an encounter with a single surface ship or a squadron of them, the submarine generally has the advantage. Despite great improvements in shipborne and airborne sonar, a submarine can conceal itself from its potential victims much more easily than they can discover it before it strikes. Nuclear-powered subs may be somewhat louder submerged than their electrically operated cousins, but other factors have more than compensated for that disadvantage: endurance that lets a submarine hide indefinitely from enemies on the surface, plentiful electricity to run power-hungry sonar and other gear, and speed enough to perhaps outrun a torpedo.

Improvements in weapons also make underwater operations safer for the submarine and more dangerous for the adversary. Smart torpedoes—the U.S. Navy's Mark 48, for example—now allow a sub to fire on a target at more than twenty times the range common to World War II-vintage tin fish. Furthermore, these weapons home on the target with sonar, greatly reducing the accuracy of aim needed to ensure a hit. Nowadays, the skipper can all but count on a hit as long as he fires his shot in the general direction of the target. After the attack, when the target is aflame and sinking and when its escorts are churning the seas bent on revenge and on protecting as-yet-undamaged vessels, the underwater maneuverability and stealth of a nuclear boat can make pursuit all but fruitless.

Long before a torpedo is fired, a submarine has an enviable advantage that aids in hearing the quarry on sonar. Sound waves in the ocean tend to bend downward toward the colder zones. Thus, much of a submerged boat's noise refracts away from surface ships trying to hear it, while the passive sonar on the sub is perfectly positioned to detect the much-louder vessels above. With today's exquisitely sensitive passive sonar, a submarine often will pick up a surface contact at ranges of twenty miles or more, farther in some circumstances *(pages 67-71)*.

Even at such distances, sonar operators can often tell a great deal about the target. They can hear, for example, the difference between a lone vessel—which tends to make a singularly rhythmic sound regardless of how many propellers it has—and the polyrhythmic sound of multiple ships. Warships' propellers, which typically turn at about seven to ten rpm for each knot of speed, sound different from those of commercial vessels, which rotate about one and a half times as fast. At a given speed, the screws of large ships turn more slowly than those of smaller craft, allowing an educated guess as to the displacement of ships approaching. How many propellers a contact has and the number of blades on each one contribute to the sounds that come from a ship in such a way that they sometimes indicate the type of vessel and perhaps its nationality. Should any of the ships be searching for submarines by means of active sonar, the frequency and pattern of pings can reveal the nationality of the ship and sometimes its class. Combining these clues with reports of the positions of both enemy and friendly vessels, a sub skipper can usually tell whether the contact is hostile. If so, he sets course to intercept the ship identified with the help of the sonar operators as the "heavy," the most attractive target of the bunch.

Passive sonar gives direction to the target, or bearing, with considerable accuracy, but every other parameter—speed, heading, and range—begins as a guess. A sonar operator can judge speed of a target within several knots by counting propeller revolutions. Of the target's heading, the sonar divulges only whether the bearing is increasing or decreasing, but any of hundreds of headings could satisfy either condition. A passive sonar says nothing about range—and thus whether a torpedo can reach the target. However, a skilled sonar operator can make a reasonable guess based on water conditions and how they affect the path that sound travels *(pages 70-71)*.

To refine the heading, bearing data from the sonar, along with the operator's estimates of the range and speed of the target, are processed by the submarine's fire-control system. In most instances, the initial result is approximate at best, with an error at least as great as those in the sonar operator's estimates of speed and range. For example, a change in bearing as the minutes pass could be achieved either by a ship that is relatively far away and plunging through the waves at a speed higher than presumed—or by a vessel that is much closer and traveling more slowly.

In order to resolve such ambiguities, the submarine traverses widely from side to side as it listens to the contact from different directions; the fire-control system, meanwhile, combines information on the submarine's speed and position from the SINS with bearing data from sonar. During an observation period lasting sometimes minutes, sometimes several hours, the system continuously shrinks the realm not only of headings, but of speeds and ranges that are consistent with the history of target bearings—as long as the original guesses were good ones. If not, the realm expands and new estimates are made to begin the process anew.

Extending the observation period increases the accuracy of this judgment, but it also exposes the submarine to greater risk of detection by ships or aircraft that are escorting the heavy. Thus many a sub skipper, sure of his crew's ability to score with a Mark 48 torpedo despite imprecise knowledge about the target's distance and speed, is satisfied with a range, for example, that may be short or long by as much as 2,000 yards, or a speed that is off by two or three knots.

Gradually the submarine approaches close enough to fire a torpedo, commonly no more than half the Mark 48's maximum range of twenty miles. Although the computers and sonar provide sufficient information to attack without coming to periscope depth,

Crewmen ease a Mark 48 torpedo down a loading sled into the USS *Chicago,* a Los Angeles-class attack submarine. Passing the 3,500-pound weapon through three decks to a berth in the torpedo room is a delicate operation that takes about twenty minutes to complete.

"under almost all circumstances," asserts Captain Gustavson, "a submariner will look. One look is worth a thousand bits of data out of your machine," a sentiment intimating that sonar readings suffer from subtle imperfections that only a pair of human eyes can discover. A glance through the periscope quickly confirms the bearings of enemy ships as reported by sonar. Seeing the target—and perhaps comparing it to ships pictured in identification manuals carried aboard the sub—is the most reliable way to authenticate it by type and even by name. Comparing the height of the ship, as measured

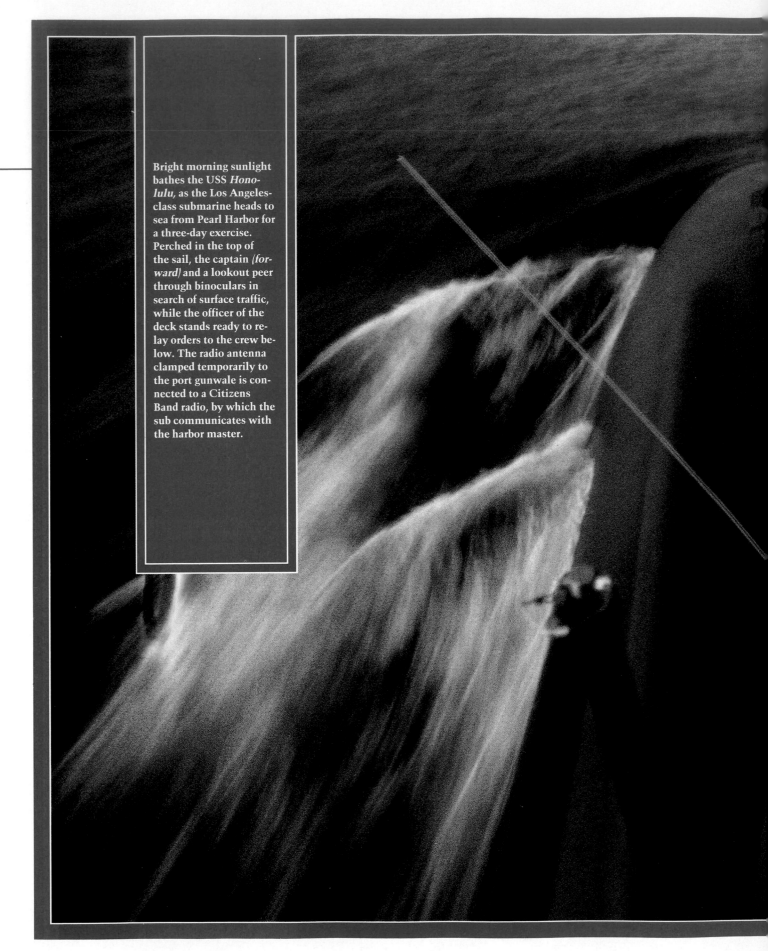

Bright morning sunlight bathes the USS *Honolulu,* as the Los Angeles-class submarine heads to sea from Pearl Harbor for a three-day exercise. Perched in the top of the sail, the captain *(forward)* and a lookout peer through binoculars in search of surface traffic, while the officer of the deck stands ready to relay orders to the crew below. The radio antenna clamped temporarily to the port gunwale is connected to a Citizens Band radio, by which the sub communicates with the harbor master.

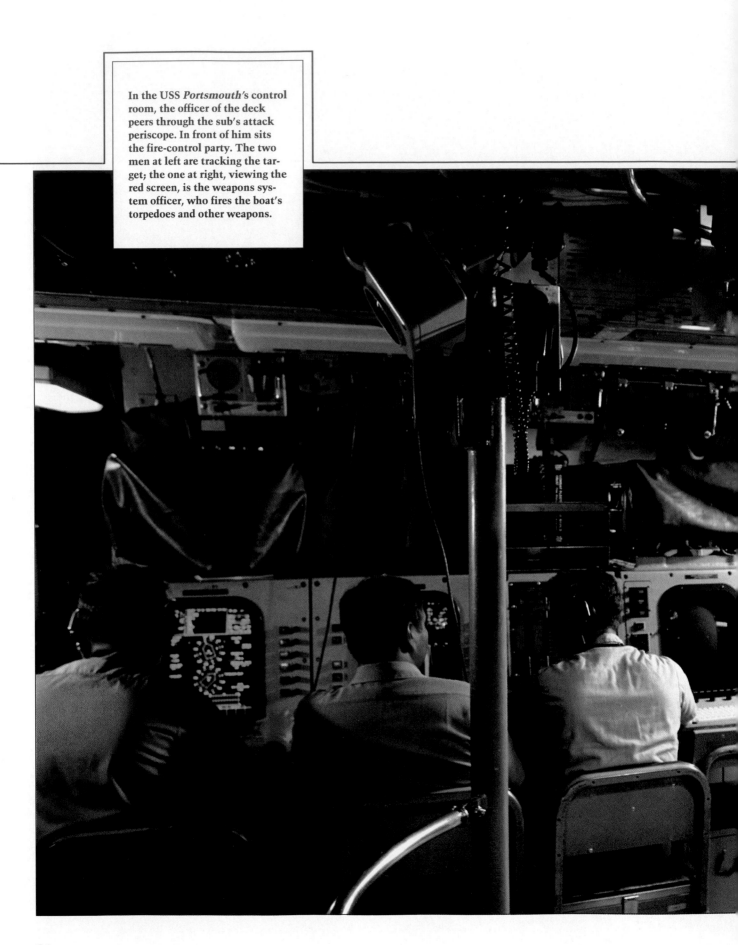

In the USS *Portsmouth*'s control room, the officer of the deck peers through the sub's attack periscope. In front of him sits the fire-control party. The two men at left are tracking the target; the one at right, viewing the red screen, is the weapons system officer, who fires the boat's torpedoes and other weapons.

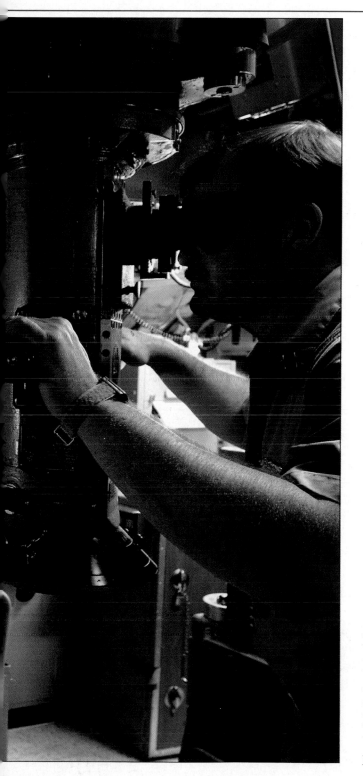

by a scale etched vertically on the optics of the periscope, with the height taken from the manual can reduce range error to about five hundred yards.

The key to remaining undiscovered while making such observations is to keep the periscope as inconspicuous as possible, even to the point, during daylight, of maneuvering so that the instrument will be unlikely to flash the quarry with a reflection of the sun. When the captain is ready for a look, he orders, "Up scope." Based on the most recent sonar bearing, the periscope operator rotates the tube so that the instrument is aimed in the direction of the target when the scope emerges from the water. By this time, the sub has slowed to a speed of only a few knots to reduce the wake produced by the periscope; the diving officer adjusts the submarine's depth so that the instrument extends no higher than necessary to see the target. The farther a periscope protrudes above the surface, the more likely it is to be spotted by an observant enemy sailor or by radar. Thus it is raised no more than a foot or so in calm seas, but as high as necessary above rougher water— up to five feet—to keep the device from being splashed by waves.

Looking through the eyepiece, the captain corrects any error in the bearing derived from sonar, then orders "Down scope." No more than ten seconds have passed, and the periscope is once again invisible underwater. According to Captain Charles Beers, former attack-sub skipper and instructor of future submarine captains at the Navy's Prospective Commanding Officer School, "there's really a personal style" to handling the periscope. Beers would shut his eyes as soon as he flipped up the handles and ordered the scope down "so I wouldn't get confused or distracted by anything else going on." The technique helped him to visualize what he had just seen as he recited the information to his tracking party. Responsible for following the progress of the target, these submariners are silent and alert in order not to distract the captain or miss a command.

For a target headed straight at the sub, one look through the periscope is enough for a successful shot with a Mark 48; other circumstances may require three or more observations to narrow the error in range, speed, and course. When satisfied that these three variables are known with accuracy sufficient for a successful attack, they are passed to the weapons-control console, which sends to the torpedo a course for it to follow.

Once the torpedo is in the water, sonar operators aboard the submarine can hear the high pitch of its whirring propellers. If the sound begins to merge with the bearing of the noise coming from the target, the chances of a hit are excellent. If the bearings do not coincide, the crew member at the weapons-control console can correct the torpedo's course by means of a toggle switch that sends steering commands to the weapon—two degrees left or right for each flick of the switch—by means of fine wires that unreel in the wake of the Mark 48.

Equipped with both passive and active sonar, the torpedo soon begins to guide itself to the mark. Some minutes into the ten-minute run, the Mark 48 typically acquires the target with the sonar operating in the passive mode. Close to the target, the torpedo's sonar goes active and guides the weapon to the kill.

Launching a torpedo is relatively noisy and can disclose the presence of a submarine to the enemy, who may detect the sound of the launch, the weapon's propellers, or the pinging of its active sonar. If they miss those clues, the detonation of a Mark 48's half-ton warhead is hard to overlook. Even then a submarine may not flee. "When you hit a couple of guys with torpedoes," says Gustavson, "and it looks like it's going to turn into mayhem up there, then you have a choice of either sticking around and participate by shooting again—or leaving. In most cases you're better off sneaking away and coming back again."

But not in every case. Should the enemy launch a torpedo in response to the attack, the sub's best defense becomes one of speed pure and simple, perhaps with a change of depth and a noisemaker decoy thrown in for good measure.

This scenario—right up to the point where the captain orders a torpedo fired—has been played out countless times by the naval forces of the two superpowers. More often than not, the surface ship never knew that it had been tracked, targeted, and "destroyed." The

As viewed through the periscope of the British Oberon-class submarine HMS *Olympus*, a Soviet Kashin-class destroyer sails in the Mediterranean. The diesel-electric sub has approached to within 500 yards of the destroyer, which carries rocket-launched depth charges, seen here angled upward just forward of the bridge.

new era in international relations may have taken some of the severity out of these rehearsals for war, but as long as both America and Russia field a force of attack submarines, the game will continue. And even though sub captains of the two navies are increasingly unlikely enemies—and if NATO is not about to fight erstwhile Warsaw Pact navies—the same cannot be said for the submariners of various other nations. The world is still riven by regional conflict, and twice in the last two decades, submarines have come into play with deadly intent.

The Indo-Pakistani War of 1971 pitted the two long-time religious antagonists against each other over East Pakistan. It was a particularly pointless war, inasmuch as the area was soon to become the independent country of Bangladesh. When hostilities broke out on December 3, an Indian task force of fifteen ships—including an ASW squadron of three frigates—sailed northwest from Bombay across the Arabian Sea to attack and cripple the Pakistani port of Karachi. En route, one of the frigates suffered a boiler breakdown, requiring it to return home escorted by the other two.

When these two frigates, the British-built 2,000-ton *Khukri* and the *Kuthar*, shaped a course to rejoin the fleet on December 9, they received a report that there was a Pakistani submarine near the Indian coast northwest of Bombay. A Sea King ASW helicopter, which was fitted with dipping sonar but not armed with torpedoes, flew out from land to assist in the search.

Unbeknown to the Indians, the new Pakistani submarine *Hangor* had already discovered the two frigates near a point of land called Diu Head. Deployed to the Arabian Sea before the war had even begun, the small, 870-ton French-built diesel-electric boat had been observing Indian naval and merchant shipping for more than a week, patiently awaiting a suitable opportunity to attack. When the two frigates steamed past, the Pakistani captain, Commander Ahmad Tasnim, had his targets.

For more than seven hours, he trailed the two frigates, keeping a safe distance so long as the helicopter orbited in the vicinity of the surface ships. Eventually, the Sea King ran low on fuel and departed. Tasnim expected a replacement, but none arrived. By now it was night, and quite cold for the area.

At this point, the *Khukri*'s Captain Mahendra Mulla decided to test an advanced sonar system that was intended to increase the range at which his ship could fire at a submarine from 2,000 yards to 20,000 yards. Like most sonar, the equipment would be ineffective at high speeds. So the captain held the *Khukri* to twelve knots, while the companion frigate *Kuthar* steamed on ahead. It is not known whether the *Khukri*'s new sonar worked properly, or whether her crew knew how to operate it correctly. It is clear, however, that the *Khukri* did not detect the *Hangor* and that Commander Tasnim slowly closed the range enough to shoot. Then, in quick succession, the Pakistani submariner launched three acoustic homing torpedoes, marking the first time that such weapons had ever been fired in combat.

On board the *Khukri*, Lieutenant Commander Manu Sharma, the ship's operations officer, had just finished his evening meal in the officers' wardroom and returned to his post on the bridge. There, he joined Captain Mulla and two other officers. Just then, the three torpedoes slammed seconds apart into the frigate's thin steel hull. One penetrated the aft magazine, a second caused flooding in the engine room, and the third torpedo demolished the wardroom Sharma had just left.

The concussion from the exploding warheads and magazine knocked Sharma briefly unconscious. When he came to, he found Captain Mulla bleeding from a head wound where he had fallen against a bulkhead. At the captain's behest, Sharma stepped outside the bridge to survey the ship. A hellish vision greeted him on the main deck. An explosion in the aft magazine had shattered the stern. Jets of flame shot from the funnel and the *Khukri* started listing heavily to starboard. Less than a minute had elapsed since the multiple torpedo hit, but the frigate obviously was doomed. Sharma made his way back to the bridge to advise Captain Mulla.

It came down to saving as many lives as possible in the little time remaining. Having been sealed against the possibility of attack by chemical or biological agents or by nuclear weapons, the ship offered only the bridge as an escape route for those trapped below, and a stream of sailors frantically scrambled up the narrow ladders to safety. Though the surrounding sea blazed with flaming oil, men hurled themselves down into the water. Suddenly the forward section of the *Khukri* lurched sharply and the bow rose up. Captain Mulla pushed Sharma overboard as the broken frigate slid below the surface, taking her commanding officer with her. The entire drama had lasted less than two minutes.

Many of the sixty or so men who made it off the ship burned to death in the fiery sea. Others drowned, sucked down with the *Khukri* when she sank. Miraculously, Lieutenant Commander Sharma survived. He spotted a dark mass in the flickering light and reckoned the Pakistani sub had surfaced to pick up prisoners. It turned out to be an empty life raft blown clear of the stricken frigate.

In a craft that was meant to hold half a dozen people, Sharma loaded 28 badly injured survivors, stacking them like cordwood, while he and a few others who had escaped unhurt clung to the side of the raft throughout the long December night. By the time an Indian ship came to the rescue eighteen hours later, only Sharma and 24 of his shipmates remained alive. The other 273 members of the *Khukri*'s crew had perished.

Commander Tasnim, meanwhile, had taken evasive action after firing his torpedoes. Going deep, he took the *Hangor* out to the open sea. Six Indian ASW ships and numerous planes hunted the *Hangor* for two days and nights. Pakistani submariners counted more than 150 depth charges, but the Indians seemed to be firing wildly without an accurate fix. Finally, the *Hangor* shook off her pursuers and emerged unscathed, her captain and crew heroes in what turned out to be their nation's losing cause.

Only one other instance of submarine torpedo combat has occurred in the years since. In the Falkland Islands War of 1982, both the British and the Argentines employed subs against enemy surface units, though with greatly dissimilar results. HMS *Conqueror*'s sinking of the Argentine cruiser *General Belgrano (pages 7-16)* made headlines around the world and strongly influenced the course of the war. Less well known are the activities of an Argentine sub that also participated in the fighting.

Argentina's navy carried four submarines in its inventory. Two were diesel-electric boats retired from the U.S. Navy and upgraded with improved batteries. One of these, the *Santa Fe*, was abandoned after being damaged by British ASW helicopters while on a supply mission to South Georgia early in the war. However, two subs were modern, German-built diesel-electrics displacing 1,440 tons and capable of twenty-two knots submerged. During trials, one had been declared too noisy and unfit for combat; it saw no action. The other, the *San Luis*, bravely sortied against the British, even though she did not approach full combat readiness. Two-thirds of the *San Luis*'s crew had been aboard less than one month. One of her four 2,400-horsepower diesel engines had broken down, slowing her surface speed and increasing the time required to recharge batteries for fully submerged cruising. Worse, the fire-control computer had checked out defective. Without a backup system, the *San Luis*'s torpedo operator would have only an imprecise idea of where to steer the German-made SST-4 wire-guided torpedoes.

Despite all handicaps, the *San Luis*'s skipper, Lieutenant Commander Fernando Azcueta, brought his boat into the war zone and prepared to make his presence known. On the night of May 1, Azcueta found a British formation of medium-size warships that he believed to be protecting a carrier. He came to this conclusion from sonar readings alone, so poor was visibility that he could see nothing of interest through the periscope. Lack of experience in penetrating an ASW screen—something his U.S. and Soviet counterparts acquired regularly—led him to conduct a standoff attack against the escorts rather than going for the possible heavy.

The *San Luis* launched one SST-4 torpedo. It left the tube and started its motor as designed, but after three minutes, the wire parted and the SST-4 continued without guidance until its fuel ran out. British ASW ships, however, had picked up the sound of its high-speed propellers and turned toward the *San Luis*. An ASW helicopter dropped at least one homing torpedo, but Azcueta maneuvered his boat to defeat it. With the element of surprise gone, the advantage had reverted from submarine to surface forces. Azcueta broke off the attack and withdrew.

Still, the Argentine skipper had plenty of fight left. On the night of May 10, Azcueta brought his submarine into San Carlos Water under the very noses of the British. Locating two destroyers on sonar and making an undetected approach, he attacked again. As before, the SST-4's guidance wire snapped short of the target. Moreover, his inexperience proved a hindrance. He had launched his torpedo with both enemy ships steaming away from the *San Luis.* Consequently, since the sub was unable to match their speed, Azcueta could not approach within torpedo range a second time and had to retire once again.

In the *San Luis*'s final encounter of the war, her passive sonar operator reported a contact that he believed to be a British submarine. Taking several bearings on the contact, Azcueta calculated the range as about 3,000 yards and launched an American Mark 37 antisubmarine torpedo at the target. The *San Luis*'s crew heard a small explosion—not large enough to be the torpedo's warhead— and the contact vanished. The nature of the explosion remains a mystery, as does the identity of the contact. In any event, no British submarine was lost during the war.

With hindsight, a submariner might rate Lieutenant Commander Azcueta's performance as one of greater determination than skill. Yet as matters later turned out, the cards had been stacked against him. When the *San Luis* returned home, it was discovered that the sub's manual steering system had been incorrectly installed. Someone at the dockyard had plugged two wires into the wrong jacks, reversing the left and right commands sent to the torpedo by the control stick. Even if the wires to the torpedoes had held, the torpedoes could not have been steered to the target.

The Unerring Flight of the Tomahawk

Only the periscopes and an antenna of the USS *Pittsburgh* broke the smooth surface of the Mediterranean Sea as the countdown reached zero hour shortly before noon on January 22, 1991, five days after the onset of the Persian Gulf War. Then a momentary bulge of water appeared forward of the periscope, and from the bulge emerged a slender, cylindrical object, leaping into the brightness on the shoulders of a flaming rocket booster. At an altitude of 1,000 feet, the rocket booster died and dropped away. Now, a small, air-breathing turbofan jet engine came to life with a soft, whistling rumble. Sprouting stubby wings, the aircraft picked up speed, banked gently around, and dropped to a height of about 300 feet above the surface of the sea. Soon it disappeared, racing at 550 miles per hour on a northeasterly course that would take it and its 1,000-pound warhead to a high-priority target many hundreds of miles away in central Iraq.

Below, in the attack center of the USS *Pittsburgh*, Commander Charles Griffiths, the ship's skipper, hung on his periscope, monitoring the launch and calling out: "Weapon broach—successful transition—continue launch sequence." Through optics in the periscope, a video camera filmed the event, and a special still camera, set to shoot six frames a second for some eight seconds, also recorded the launch.

Air-breathing, sub-launched cruise missiles are not new. Shortly after World War II, the United States experimented widely with the 560-mile-range Regulus as a land-attack weapon before shelving the program in 1964 because of the introduction of the Polaris ballistic missile to the U.S. submarine fleet. As for the Soviets, they had also embarked on a massive cruise missile program. By the late 1980s, the Red Fleet boasted no fewer than sixty-eight nuclear-powered cruise-missile subs (SSGNs) in eight classes. One of these boats, a converted Yankee-class boomer, was equipped to launch a dozen SS-NX-24 nuclear-armed cruise missiles to a range of 2,200 nautical miles. But the rest of the Soviet SSGN fleet concentrated on naval warfare, with an array of weapons capable of attacking surface ships and submarines at distances ranging from twenty nautical miles to twelve times as far. The most formidable of the SSGNs, six huge Oscar-class boats, each could carry two dozen cruise missiles tipped with either nuclear or high-explosive warheads, plus an equal number of torpedoes.

America had nothing comparable in its naval arsenal, but meanwhile, Navy weapons planners had come up with a better idea: the Tomahawk family of land-attack missiles (TLAMs) and antiship missiles (TASMs), an amazingly versatile weapon that was designed to fit inside a standard torpedo tube. Thus the weapon could be launched not only from the air, land, or surface of the sea, but also from beneath the waves. By 1991, Tomahawks were operational on 75 percent of the ninety-nine-ship U.S. attack fleet, with another sixteen vessels abuilding.

A Los Angeles-class submarine could carry a dozen Tomahawks, displacing an equal number of torpedoes. About one-third of the

class had been fitted with a vertical launch system of twelve tubes installed forward of the sail, permitting these ships to carry Tomahawks without sacrificing any torpedo load. And what awesome weapons they are! In its nuclear version, which was withdrawn to storage in a reduction of tactical nuclear weapons begun in mid-1991, the Tomahawk could deliver a 200-kiloton warhead 1,400 nautical miles with a 50 percent chance of striking within 250 feet of intended ground zero. The high-explosive warhead of the shorter-range conventional model, called the TLAM-C, routinely detonates within 30 feet of the aiming point.

When the *Pittsburgh*'s target assignments were first radioed to the ship twelve hours before launch on that sunny January morning, Griffiths assembled his key weapons personnel and watch standers in the submarine's wardroom. For the next several hours they huddled over the launch plans. Griffiths and his team plotted firing and target coordinates, distance and range, courses and way points.

As the launch approached, they evaluated the various other elements that would influence the Tomahawks' flight: predicted sea state, ambient air and water temperature, wind direction and force. According to Griffiths, the Tomahawk is a "robust system. It can be fired in significantly degraded weather conditions." But it works best launched into the wind and as close to the vertical plane as possible; too much roll, caused by violent weather on the surface, will force a postponement. Once aloft, the missile can overcome most adverse conditions, such as heavy precipitation. But extremely hot temperatures—120 degrees and up—will affect engine output and thus terrain-following performance and range. Griffiths and his launch team considered all this, made their computations, and fed the masses of data into the Tomahawks' guidance systems.

Although Griffiths could fire Tomahawks from deep beneath the

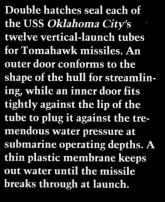

Double hatches seal each of the USS *Oklahoma City*'s twelve vertical-launch tubes for Tomahawk missiles. An outer door conforms to the shape of the hull for streamlining, while an inner door fits tightly against the lip of the tube to plug it against the tremendous water pressure at submarine operating depths. A thin plastic membrane keeps out water until the missile breaks through at launch.

Less than twenty seconds pass between the instant that the launch officer aboard a Los Angeles-class sub punches the button to dispatch a Tomahawk and the time that the missile disappears from view as the skipper watches through the periscope. Although the crew cannot influence the Tomahawk's flight, the captain observes the launch to

A Tomahawk broaches.

Rocket exhaust flares below the missile.

confirm that the weapon leaves the water with the booster lit. If not, another missile may be sent toward the same target.

As launch time neared, "we were all business," says Captain Griffiths of the *Pittsburgh*, a sub that fired Tomahawks during the Gulf War. "I wanted to make sure there was no amateur hour. But once things were over, we felt back-slapping good." With a camera attached to the periscope, he took these photos of a missile leaving the water as a memento of the event for his crew.

The wing cover pops off.   Pieces shrouding the tail fins fall away.

surface, this mission called for dispatching the missiles from a depth of sixty feet to avoid accidentally hitting a passing merchantman. "Surface shipping was heavy," said Griffiths. "We had to periodically maneuver because there was completely oblivious merchant traffic in the area." Submariners dislike operating for any length of time at periscope depth, but Griffiths had few worries on this occasion. The *Pittsburgh* was operating in friendly Mediterranean waters, accompanied by U.S. surface ships.

Although the missiles can be launched with as little as a few hours' notice, Griffiths began his countdown twelve hours before launch time; having never fired a load of the missiles in combat, he wanted to be certain nothing went awry. Early hours were spent rechecking everything, from ship's systems to surface weather, and preparing contingency plans for use in case one of the missiles malfunctioned or the submarine somehow came under attack. Six hours before launch, he assembled a special battle team of those who would be directly involved with firing the Tomahawks, as well as sonarmen and radio operators.

With two hours to go, Griffiths called the rest of the *Pittsburgh*'s crew to battle stations—and that meant everybody on the boat. "There wouldn't be anybody sleeping," he said wryly. "You'd have damage-control parties stationed to cover contingencies. They might be sitting in the crew's mess drinking coffee—but with special damage equipment right next to them, to react in case you have a system fire or something."

Senior people manned key posts in the engineering, electronics, and command spaces. The most seasoned diving officer checked the ballast controls while the helmsman and planesman manipulated the ship's controls to keep the *Pittsburgh* in perfect trim so that depth could be regulated precisely at any speed. Griffiths could not simply bring his boat up to the launch coordinates and hang there, waiting for the time to launch. A submarine cannot hover effectively without special thrusters, so Griffiths maintained just enough way for control and eased up to his launch point in a feat of precision navigation that would put him at exactly the correct spot at exactly the right second.

Minutes before launch, Griffiths initiated the final countdown sequence. One of the twelve missile hatches forward of the *Pittsburgh*'s sail opened. Griffiths stood at the periscope as the count reached zero, then the boat lurched slightly and a whooshing roar could be heard as compressed gas propelled the first Tomahawk out of its tube and seawater rushed in to replace it. Then another missile entered the last seconds of countdown.

If called upon, the *Pittsburgh* and her sister ships could launch a Tomahawk every ten seconds, all twelve in less than two minutes. But the Persian Gulf War demanded a less intense salvo; the Tomahawks played a smaller role than they might have because they had never been tested in combat, where many things can go awry and results count. All told, the *Pittsburgh* and another Los Angeles-class submarine in the Red Sea, the *Louisville*, together fired twelve Tomahawks.

The crews learned later that they and their Tomahawks had performed excellently. The missiles had hit with a great degree of accuracy, and the success of a new weapon entering the inventory was highly gratifying. Yet, as in minelaying, there remained something missing for dyed-in-the-wool members of the silent service. "You know," said Griffiths, "it's not as satisfying to see a missile fly over the horizon. A submariner would like to be much closer, more intimately involved in the end game—and see the destroyer split in two through the periscope."

Tomahawk missiles are certain to have a profound influence on the way attack boats operate against surface targets at sea as well as on land. Yet there is one arena where the submariner remains personally and intimately involved. And that is beneath the waves, in the depths of the ocean, when a submarine is sent to hunt down and kill another submarine. ★

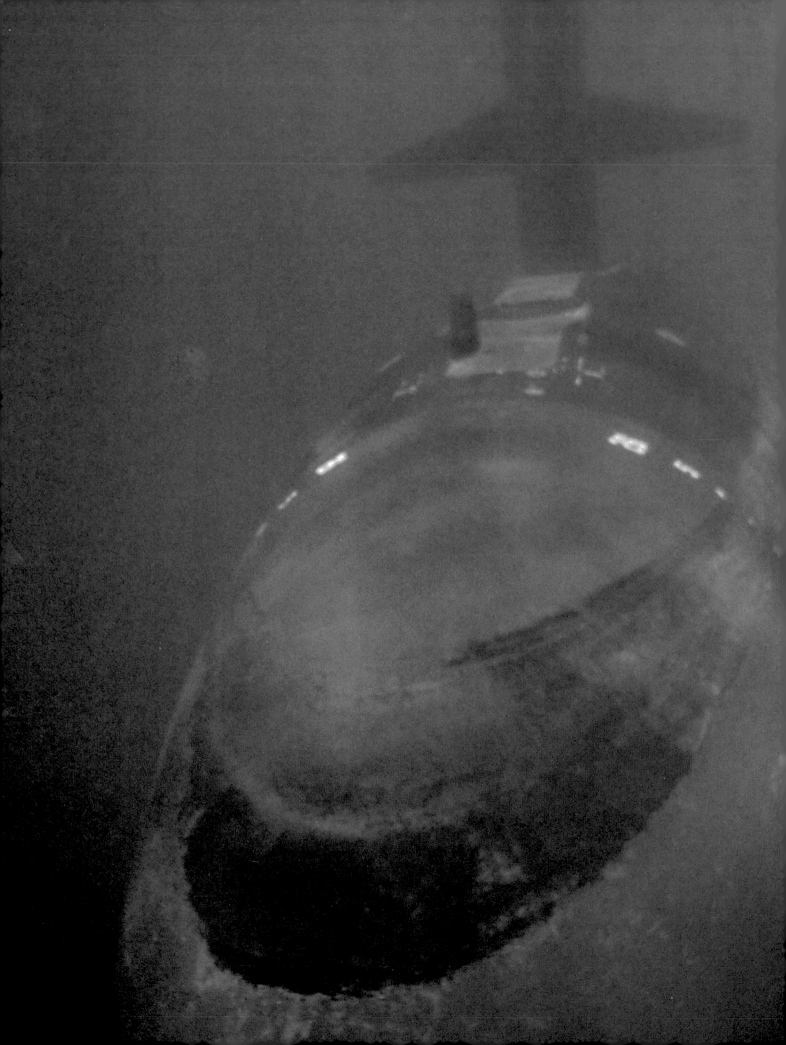

4

War of the Silences

Lurking just 70 feet or so below the waves, the USS *Cavalla,* a Sturgeon-class attack submarine, passes almost unseen through the clear waters of the Atlantic Ocean off the Bahamas. Were the sub 100 feet deeper, it would be completely invisible. The dark stain curving around the bow is algae.

Sitting in a dimly lighted, narrow alcove just forward of the control room of the USS *Tautog,* a nuclear-powered Sturgeon-class fast attack submarine, a young sonar operator pressed his headset tight against his ears and leaned forward with intense concentration. For hours on this June day in 1970, he had been monitoring the *Tautog*'s sensitive BQS-6 sonar, a spherical array of hydrophones mounted in the bow. Through this extraordinary set of electronic ears, the sonarman could hear the sounds of a Soviet submarine—a 5,200-ton Echo II-class guided-missile boat running a few hundred feet beneath the surface of the northern Pacific, several hundred yards ahead of the 4,250-ton American ship.

Built into the upper surface of the Echo's 380-foot-long hull wcrc launchers for eight SS-N-3A Shaddock antiship missiles, which the submarine could fire only from the surface. Immediately aft of each launcher, the boat's hull formed a depression. Slope-sided and several feet deep, the notches were designed to deflect the missiles' flaming exhaust away from the hull, but to the Soviets' chagrin, they did more than that: When the Echo submerged, the deflectors roiled the stream of water flowing along the hull, making a terrific racket that grew louder the faster the boat surged through the water. Listening aboard the *Tautog,* the sonarman used the loudness of this noise and the sound of the Soviet vessel's turning propellers to estimate the Echo's range, which, along with the sub's bearing, he regularly reported to the control room.

There one of the members of the tracking party deftly plotted target bearings on a long sheet of white paper spread across a chest-high table just aft of the ship's observation and attack periscopes. Each time, he reached for one of the pencils taped to an overhead lamp, different colors for different contacts. Then he slid a protractor across the sheet, rotated it to match the bearing reported by sonar, and drew a straight line radiating from a tiny, glowing *X* that marked the *Tautog*'s location. The symbol was projected onto the underside of the paper by a light, called a bug, that is synchronized with the ship's inertial navigation system and moves in scale with the ship's speed and course. Over time, the sets of such lines, called the plot, yield a bounty of information, including two specific measures: how much the target bearing changed from reading to reading, and the rate at which it changed. Plugging them into a computer programmed with a surprisingly simple equation quickly calculated a rough estimate of the Soviet ship's range. Distance between the two boats led to other computations that yielded depth.

Analyzing the plot, the tracking party noticed that the Echo had veered from the course it had been following for some time. The commanding officer of the *Tautog* responded prudently. Seeking a safe place to avoid detection until the Soviets' motives became clear, he ordered his sub to a greater depth and had it come to a virtual standstill, a trick that allowed the ever-present murmur of the ocean to drown out the weak sound of his boat's propeller.

As soon as the *Tautog* leveled off at the new depth, word came from the tracking party that the Soviet vessel was executing a long, arching turn that quickly reversed the Echo on its course and sent it running right down its own wake, more or less toward the *Tautog*. The Soviet sub was compensating for a weakness shared by all submarines—the inability of bow-mounted sonar to hear anything behind the ship, the region known as its baffles. For commanders of early American attack subs, which, like their Soviet counterparts, had no sonar to listen behind the boat, a slow turn of 60 to 120 degrees usually made sounds in the baffles audible. Soviet skippers, perhaps supplied with less capable sensors in the bow, felt compelled to turn completely around. "An advantage of the maneuver," explains Commander Richard Pariseau, a retired submarine officer, "is that it leads you to think they may have detected you so you are tempted to move and give yourself away." American submariners call the maneuver the Crazy Ivan.

But the *Tautog*, masked in silence a safe distance below the Echo, did not spook, and the Soviets failed to detect it. After a while, the Soviet ship turned and resumed its original course, much to the relief of the 107 men marking time aboard the American sub. Their complacency, however, would prove to be short-lived.

An hour or so later, the Echo turned once again—apparently to clear its baffles a second time. As before, the *Tautog* maneuvered to minimize the chance of being detected, but this was not to be an everyday Crazy Ivan. Hunched over the plotting table, an analyst studying the bearings reported by sonar stated gravely that there was no change from reading to reading; the Echo was on a constant bearing. If the two submarines continued on their present courses, a collision was inevitable.

Immediately, the commander ordered a steeper dive, but the Echo kept charging. Listening as the sound of the Echo's propeller and the noise of water streaming along the hull grew progressively louder, the sonarman estimated the Soviets' range and reported: "Contact at 500 yards and closing fast."

Soon, the clamor of the approaching sub became audible to everyone aboard. They could only hope that the *Tautog* had dived deep enough in time to avoid collision. But no. Metal screeched against metal, and the American ship rocked violently. Officers and men on watch were slammed against bulkheads, and off-duty crewmen were flung from their bunks. Hand tools that had been stored in lockers broke loose and ricocheted around the submarine's compartments. Miraculously, no one suffered a serious injury, and the *Tautog* sprang no leaks.

In the sonar room, the sailors listened intently for signs that the Echo had also survived the collision. At first, they heard not a sound. But then they picked up an eerie succession of sharp reports—one sonarman said they sounded "like the popping of popcorn"—followed by a cryptlike silence.

Puzzled, the commander of the *Tautog* had his ship reverse course. "We turned completely around and just listened—nothing," said a member of the submarine's crew. "Then we came up slowly. We listened and listened again. Nothing. We came up to periscope depth and looked and waited. Nothing." After half a day, the *Tautog* gave up the search.

Navy investigators later concluded that the two subs had probably come within a heartbeat of escaping the collision. The Echo's

Shrouded by fog, the USS *Whale* slips out of the submarine base at Groton, Connecticut, and heads for deep water. Protruding vertically behind the two observers standing watch in the sail of the boat are radar and radio antennas and a periscope, used by the navigator when the sub is on the surface.

hull had cleared the *Tautog* when one of the missile ship's twin propellers clipped the top of the American sub's sail. The impact likely ruptured the seal surrounding the Soviet submarine's propeller shaft, allowing thousands of gallons of seawater to spurt into the engine room. As the compartment became flooded, the ship started a horrifying free fall. Somewhere below 1,400 feet, the Echo's hull collapsed loudly.

This rare mishap, in which about a hundred Soviet seamen lost their lives, was the outcome of a deadly serious game that American and Soviet submarines have played every day with each other since the 1950s. The contest originated during the tensest years of the Cold War, when the United States assigned three missions to its fleet of attack submarines. Fears of a Soviet assault on Europe spurred the first two: defending the aircraft carrier battle groups that would sortie against enemy coastal installations and keeping open the tenuous sea lanes of communication by which America would resupply her own troops and those of her NATO allies. Toward these ends, the early Skipjack and Permit classes, and later Sturgeon- and Los Angeles-class subs, were to surge deep into the frigid waters of the Norwegian and Barents seas to track down and kill Soviet attack submarines before they could escape to their hunting grounds in the wide Atlantic.

The attack subs' third mission was keeping tabs on the huge nuclear-powered submarines that the Soviets, like the Americans, began using as launch platforms for ballistic missiles. At first, the noisiness of the early boomers and the limited range of their weapons made the U.S. Navy's task relatively easy: In order to threaten targets in North America, the Soviet ships had to exit their home waters into the Atlantic and Pacific oceans, and they could reach them only by sailing through choke points that were monitored by fixed undersea listening devices and patrolled by NATO surface ships and aircraft. Cued by these assets, the American attack subs were usually able to start shadowing the boomers before they reached the open ocean.

Over the years, however, the range of submarine-launched ballistic missiles grew until it came to rival that of the land-based variety. The enhanced striking distance enabled the boomers to remain in waters close to the Soviet mainland, where the big subs could be hidden by a ceiling of thick ice and protected by a defensive screen of smaller attack subs.

In adopting this tactic the Soviets reduced the number of potential threats to their boomers to just one: the American attack submarine. Of all the U.S. Navy's assets, only subsurface hunters such as the *Tautog* had a chance of finding the Soviet boomers in their Arctic bastions and sinking them, if necessary, before they had time to move to an opening in the ice, surface, and loose their weapons. And the situation is no different today, even after the thawing of the Cold War.

Charting a narrow course between safeguarding the peace and provoking a conflict, American Los Angeles-class subs still regularly slip under the icecap—not because they think the Soviets will launch their missiles anytime soon, but because the boomers remain capable of firing them. And as long as this holds true, American submariners will likely be sent to find the missile ships and to gather as much information as possible about the sounds they make, the places they hide, the routines they follow, and the attack subs that defend them.

Keeping in Touch

Deep beneath the wind-whipped surface of the Atlantic Ocean, a lone Los Angeles-class attack submarine runs at more than thirty knots and then, for no apparent reason, suddenly slows. Inside, the commanding officer has ordered, "All stop." The 360-foot-long vessel coasts through the water.

Several hours have passed since he last tuned in to a signal from one of the Fleet Satellite Communications (FLTSATCOM) System's several orbiting radio-relay stations that the Navy uses to keep in touch with its wide-ranging attack submarines. The satellites—the same ones the president and the secretary of defense rely on to communicate with field commanders around the world—are positioned in geosynchronous orbits 22,300 miles above the ocean waves. At regularly scheduled intervals, a number of times every day, the satellites pass messages from shore-based command centers to the boats by means of ultrahigh- and superhigh-frequency (UHF and SHF) radio transmissions, sent in short bursts to make them difficult to intercept.

Most of the time, the messages pertain to the administrative business of fleets the world over—resupply schedules, crew rotations, and other logistical matters. Every couple of weeks, a crewman might receive a thirty-six-word personal message from a family member. But from time to time, a submarine receives a message with new orders.

There could be any number of reasons for such a change: Perhaps an orbiting reconnaissance satellite has gathered evidence of a new type of sub putting to sea for a maiden voyage. During the Cold War, such spacecraft took high-resolution photographs of Soviet submarine bases to determine which boats were at sea and which in port. Occasionally, the camera caught a submarine cruising on the surface of the shallow inlets that lie between their moorings and deep water. The top-secret pictures were so detailed, quips one analyst, that "you could tell if the guys on the bridge watch had their parka hoods up."

Inside the USS *Atlanta*'s sonar room, specialists wearing headphones monitor ocean sounds picked up by the boat's listening apparatus. Arrayed in front of the sonarmen are video screens showing graphic displays of the noises (page 42) as well as control panels for operating the sonar. Dim blue light in the sonar room minimizes eyestrain.

Attack subs have the option of not receiving every broadcast. Unable at times to approach the surface without the risk of being detected, or unwilling to break sonar contact with a submerged target, they can always call up the satellite whenever they choose and command it to dump data. But if a sub commander waits too long before signaling the spacecraft, he might find that his orders have changed drastically in the meantime and that his ship is now out of position, behind schedule, or worse. To prevent this, most subs plan to copy a regularly scheduled transmission once every eighteen to twenty-four hours, when the Navy updates the batch of messages that the satellites relay.

Eager to do this, the submarine commander, having slowed his vessel, instructs his crew to begin the elaborate process of bringing the boat shallow enough to raise a radio-receiver antenna above the waves. The first step on the way up: listening with passive sonar, including the BQR-23 array, a bundle of hundreds of hydrophones, three inches in diameter, that the submarine tows behind it at the end of a long, slender cable about the diameter of a sailor's little finger and more than seven times the 360-foot length of the ship. The cable protrudes from an opening in the upper star-

board surface of the vessel's hull, forward of the sail, and extends into the submarine's wake. Sonarmen refer to the array as their "rearview mirror" because it enables them to hear sounds that originate in the sub's baffles and elsewhere in the hemisphere behind the boat.

The effectiveness of the BQR-23, like all passive arrays, decreases when the sub travels fast; the cable vibrates in the boat's wake, making a hum that masks the sounds that sonarmen listen for. When the sub slows to less than fifteen or twenty knots, however, the cable is quiet, and sounds come through loud and clear.

After the sonar operators satisfy the captain that they hear nothing that suggests a risk either of being detected or of colliding with surface traffic, the commander has the boat come to a depth of 150 feet. Then he orders the ascent halted and, though the sub is still far below the surface, has the periscope raised. Not only can the head rotate to every point of the compass, but it can pivot to look almost straight up. In the light of day or a full moon, this feature gives the skipper an opportunity to look for dark shapes on the surface—hulls of ships that sonar somehow missed.

"The most dangerous spot for a submarine is between periscope depth and 150 feet," observes Pariseau. "You want either to have a periscope up or to be deep enough so that you don't get hit by these deep-draft ships such as oil tankers." At a depth of 150 feet, measured to the keel, the periscope head of a Los Angeles-class sub rises to within 85 feet of the surface, near enough to be struck by a loaded tanker. When headed directly toward a submarine, these huge ships often remain undetected by sonar. Not only is the roar of the hull passing through the water indistinguishable from many sounds of the ocean—a storm-tossed surface, for example—but it is so loud that it seems to come from all directions at once and masks the sound of the tanker's propellers.

Avoiding a collision in instances such as these, although they are rare, requires quick thinking and perfect teamwork on the part of the men inside the submarine's control room. " 'Emergency deep.' That's all the commander has to say," asserts Captain Fred P.

Gustavson, the former Sturgeon-class skipper. "Everybody on board knows that they're to take set actions: The scope comes down, the planes go to dive, and speed is increased." And as the trim tanks flood, the submarine drives down quickly, putting a layer of safety between itself and the tanker.

"If the captain sees nothing above his boat, he orders it to periscope depth as he continues to look for shadows above. Then with the periscope head right above the water line," says Gustavson, "you make about three quick sweeps all the way around just to make sure that there's nobody there." During the first sweep, the skipper looks for any surface ships that might be nearby. During the second, executed with the head of the periscope tilted upward, he scans for aircraft. In the third, he scours the horizon for long-range surface contacts.

The procedure seems foolproof, yet surprises occur. "One time in the Med," says Captain Charles Beers, remembering his tour as an executive officer aboard a Los Angeles-class sub, "the captain called me up to look. It was a moonlit night, and right in front of us— probably a couple thousand yards away—was a sailboat, just between us and the moon." Powered only by the wind, the small craft cut noiselessly though the water. "The only reason we knew was because we were looking through the periscope."

As the skipper peers through the periscope, a detector mounted atop the instrument listens for radar signals from aircraft—or from a ship too distant to see. Modern surface-search radar can spot a raised periscope at ranges as great as forty nautical miles.

During a drill, the USS *Buffalo* broaches whalelike after an emergency ascent from a depth of some 200 feet. The maneuver, which is accomplished by forcing all the water from the main ballast tanks with compressed air (page 19), takes about one minute from such a depth and is reserved for dire situations such as a serious fire or a leak.

If no vessels can be seen and no radar emissions detected, the captain orders the electronic countermeasures mast extended; at full height, it reaches less than five feet above the surface of the water. One of twelve pieces of equipment that can be run up from the sail of a Los Angeles-class boat—including an identification beacon, a snorkel, several radio antennas, and the sub's own rarely used air- and surface-search radars—the countermeasures mast bears a specialized antenna that intercepts all manner of radar waves and passes them to an analyzer that helps the crew to classify radars as hostile or friendly according to the pulses they emit. Only under such vigilance is the submarine at last ready to copy a satellite communication.

The broadcast typically arrives in an encrypted UHF or SHF burst

Glistening after an abrasive shower of sand and steam, the *Billfish* sits in dry dock at Groton. Applied when a sub is pulled from the water for repair or alteration—every four years, on average—the treatment removes old antifouling paint as well as algae, barnacles, or other growths that reduce speed.

that is decoded automatically on board the boat and printed out instantly on a teletype machine. The signal begins with an index that notifies the submariners of any dispatches that are addressed to their boat. "If there's a new message for me, I'll wait and copy it," says Commander Pariseau. "But if I already have them all, it's all over. I'm done."

Playing a Team Sport

The time-honored role for the attack submarine—and the one preferred by its crew—is that of the lone wolf, top dog in its own territory. On many occasions, however, these man-made predators of the deep are called upon to join forces with others of their kind or with a surface task force.

Often, a joint operation begins with a rendezvous, and almost always, a submarine en route to such a meeting travels fast and deep, deaf and blind. Nevertheless, the skipper can find his way unerringly. An inertial navigation system is the key; the one aboard Los Angeles-class subs can track the boat's position to within tens of meters. However, errors in the system accumulate over time, and maintaining such precision requires that the device be given an accurate fix on the submarine's location every couple of weeks or so, something that can be accomplished whenever the vessel comes to periscope depth. Among the special antennas that extend from the sail is one that receives signals from the Global Positioning System, a constellation of satellites that gives a position accurate to less than ten meters.

Accuracy in undersea navigation is a prerequisite for many tasks that an attack submarine might be called on to perform, such as shielding an aircraft carrier battle group from underwater adversaries capable of sinking the flattop by missile or torpedo. This responsibility, of course, does not fall on submarines alone—or even primarily. A carrier battle group has other antisubmarine resources, and they are considerable. Surface ships such as the 8,040-ton Spruance-class destroyer and 4,250-ton Knox-class frigate not only tow passive sonar arrays that can detect submerged targets under ideal conditions at ranges as great as 100 miles but can also dispatch sonar-equipped helicopters armed with torpedoes ahead of the battle group. Working in concert with maritime patrol planes such as

the land-based P-3 Orion and the carrier-launched S-3 Viking, the choppers give the battle group the ability to strike quickly and at great distances. But of all a battle group's ASW assets, only an attack submarine has the endurance to stay on station indefinitely, in bad weather as well as good. Furthermore, only a sub has the ability to listen for enemy boats screened from surface sonar by temperature differences in the water *(pages 70-71)*.

To help prevent the tragic loss of a submarine to friendly fire, from either above or below the surface, a battle group commander assigns the submarines in his service to well-defined search areas. Often, they are established ahead of the carrier and across the most likely axis of approach for enemy subs. But depending on the situation, the underwater ASW screen can also be positioned along the flanks of the battle group, or behind it, where they can watch for intruders that might try to sneak in the back door.

The size of the areas depends not only on the range of the subs' weapons and sensors—which can cover an area of a couple of thousand square miles—but also on the number of submarines available for the job and on an educated guess of where the hostile subs are most likely to come from. "You would establish a search area and move back and forth on a pattern perpendicular to the transit lane, looking for the enemy to come," explains Pariseau. Working far in advance of the fleet, "you would do your searches in a carefully orchestrated way—similar to the way you cut your lawn—back and forth, keeping a good plot of every turn." Closer in, the sub might sprint from place to place within the search area, pausing briefly to listen at each stop.

Nearby, a second and maybe a third submarine do the same. Streaming towed arrays, the ships diligently search within discrete operating areas that are most often aligned end to end across the threat axis. Each area shares a boundary with the one adjacent to it, but there is little risk of conflict. According to strict rules of engagement promulgated beforehand, the subs are prohibited from attacking across borderlines, and sonar operators on each ship can identify the acoustic signature of an American submarine as friendly, reducing to almost nil the possibility of one Los Angeles-class boat, for example, torpedoing another. Collision is equally improbable; inside the control room of every submarine, crewmen monitoring the inertial navigation system give ample warning whenever their boat nears a borderline.

"The attack subs are there," notes Pariseau, "to detect enemy submarines—particularly missile shooters, which can strike at ranges far greater than those carrying only torpedoes—and first warn the carrier." Notifying the carrier of an approaching hostile sub poses a vexing problem: how to send the warning without losing contact with the target or drawing unwelcome attention from other enemy forces. Submariners are reluctant to rise to periscope depth; doing so makes noise that the target submarine might use to turn the tables. Additionally, raising a radio mast offers a target for the enemy to see on radar and might permit eavesdroppers for the opposition to pick up the signal and trace it to the source.

The solution can be found on the deck immediately below the control room of the submarine. In a cramped compartment that doubles as the medic's office, a seaman loads a yard-long buoy, three inches in diameter, into the boat's signal ejector, a device that operates somewhat like a torpedo tube *(pages 144-145)*. After flooding and pressurizing the tube, he pushes a button, kicking the seven-pound cylinder into the ocean. Used for many other communications tasks besides transmitting a warning, the buoy climbs rapidly and steadily to the surface, where it floats in an upright position, and an antenna extends out of the end that bobs above the water line. Containing a tiny audiotape player, the buoy also has a timer and a VHF transmitter. Together, these components broadcast a warning recorded just before the buoy is released. Encrypted to prevent the enemy from learning that he has been discovered, the message includes the type of submarine detected and its location, among other details.

The appeal of the buoy to submariners resides chiefly in its timer, which can be set to delay the message up to an hour. During that time, the submarine can move twenty miles or more from the point of transmission, much reducing the likelihood that enemy forces intercepting the signal can find their way directly to the boat. Picked up by a friendly surface ship or aircraft, the message is relayed to the admiral in charge of the battle group. Usually, the submarine then moves in for the kill.

In the event of multiple contacts or a malfunction aboard the boat that prevents it from pressing the attack, the battle group commander will detail one of his ships or aircraft to find the intruder and sink it. Under these circumstances, the picket submarine runs a risk of becoming the target. Though the hazard is reduced by de-

War Games in Neptune's Realm

From Andros Island in the Bahamas, the U.S. Navy operates an underwater arena called the Atlantic Undersea Test and Evaluation Center (AUTEC). The location was chosen for a hydrographic feature known as the Tongue of the Ocean. This hundred-mile-long channel, up to 3,400 feet deep, is almost completely surrounded by reefs and islands that largely fence out commercial vessels. On two ranges in the tongue, submarines, surface ships, and aircraft fight mock battles that test the mettle of submariners and antisubmariners alike.

Charged with keeping track of these participants is the AUTEC Real-Time Graphic Operating System, or ARGOS. It consists, in part, of hydrophones that track surface vessels, submarines, and dummy torpedoes fired on the ranges. To do so, the hydrophones pick up signals from special "pingers" fitted to each participant but inaudible to them. Half a dozen radars follow aircraft. Data from all these sensors passes to computers ashore that display the action on video terminals as it happens (right).

Exercising at the AUTEC, submarines can stalk each other or surface vessels even as the prey seeks to evade or sink them. To improve combat skills, any competitor can fire warheadless weapons at artificial submarines—specially outfitted torpedoes that sound like subs or surface ships to sonar. And the entire skirmish can be recorded for later review, helping the warriors to profit from errors the next time they sally forth.

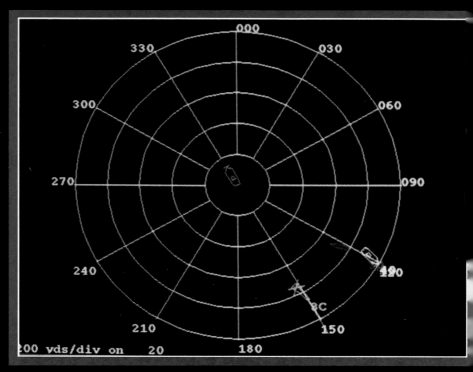

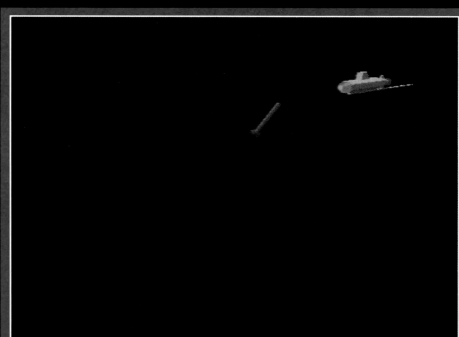

The Navy is secretive about real drills at AU-TEC, but the computer simulation at left shows how an encounter appears to system operators—and to players reviewing the game. In this instance, a submarine (yellow) fires a torpedo (red) at another sub (orange) as a helicopter (green) waits to retrieve the dummy weapon. The upper panel shows an overhead view of the situation about one minute after a torpedo has been fired at the target sub, located at the center of the circle; range rings are 1,200 yards apart. The same situation, as seen from a position near the attacking submarine, appears in the lower panel. Both views reveal that the torpedo does not head straight for the target, in part to conceal the attacking submarine's location.

laying the warning while the boat withdraws, a sub might also set a particular course or cruise at a speed that friendly ships and aircraft, following rules of engagement, recognize as a sign of a friend.

Most likely, however, the submarine will turn immediately after ejecting the buoy and head for a NOTACK, or no-attack, zone—a small, rigidly defined haven. According to the rules of engagement, surface ships and aircraft that detect a submerged target in a NO-TACK zone are to continue the search elsewhere.

For all that attack submarines can contribute to the defense of the surface navy, the role is a secondary one. Their primary mission since the invention of the ballistic-missile submarine has been to seek out these boats wherever they might hide, in the hope of sinking them before they can loose their terrible rain of nuclear warheads on the United States or its allies.

Stalking a Boomer

Almost seven years after the Cuban missile crisis of 1962, Soviet rockets were again poised within a few minutes' flight of the United States, this time aboard submarines. The USS *Lapon*, a Sturgeon-class attack ship, was prowling silently beneath the choppy surface of the Atlantic Ocean, not far off the eastern seaboard of the United States. Resting deep in the sub's steel bowels was a formidable array of maritime firepower: four Harpoon missiles, good against surface ships as far away as eighty nautical miles; fifteen wire-guided Mark 37 torpedoes—precursors to the Mark 48 torpedoes that Los Angeles-class submarines carry today—for sinking both surface and submerged targets at ranges up to five miles; and four SUBROC antisubmarine rockets, the only weapons aboard the *Lapon* bearing nuclear warheads.

The twenty-one-foot-long SUBROC had a range of about twenty-five nautical miles. Like the Harpoon and the Mark 37, it was designed to be ejected from the attack sub's torpedo tubes. After launch, a solid-fuel rocket motor would ignite and power the 4,000-pound missile up through the water and into the air. The missile would streak toward the target under inertial guidance until the booster burned out. Then it and the warhead—a nuclear depth bomb—would separate, and the munition would fall back beneath the waves, sink to a preset depth, and detonate, unleashing a violent underwater shock wave almost certain to crush any sub unlucky enough to be lurking within five miles.

Commander of the *Lapon* was Captain Chester "Whitey" Mack, a six-foot-five-inch former college basketball player who had gained esteem within the silent service in the mid-1960s by outrunning a torpedo that homed on his sub when he snooped too close to a Soviet naval exercise. Striding into the control room, Mack cocked his head to avoid a low-hanging lamp, took a quick, hard look at the digital clock mounted over the pair of plotting tables, and ordered the officer of the deck to cut the boat's speed. The Soviet boomer they were tracking, Mack predicted, would be slowing soon.

Churning through the sea far ahead of the *Lapon* was the largest undersea craft yet constructed in the USSR—a 426-foot-long, 7,900-ton Yankee-class submarine carrying sixteen SS-N-6 ballistic missiles. The monster sub's size befitted its ominous capabilities: From a location just 700 nautical miles east of New York City, the boomer could shower one-megaton nuclear warheads down on targets as far west as New Orleans, St. Louis, and Minneapolis. And it was reported that the Yankee required extremely little time to do so: From just beneath the surface, the boomer could fire all its single-stage, liquid-fuel SS-N-6s in less than two minutes.

How the *Lapon* had latched onto the Yankee remains secret, but it is far from unlikely that the boomer was first detected by SOSUS, the Sound Surveillance System, shortly after it left home waters on its present mission. SOSUS was initially constructed off the east coast of the United States in the 1950s and then at thirty-five other locations around the globe where Soviet ballistic-missile submarines passed en route to launch zones. The system comprises hundreds of hydrophones laid across the ocean floor that can tell within fifty miles the location of a submarine, even if it is several thousand miles away from the nearest SOSUS array.

Sound that is gathered by the hydrophones is transmitted through undersea cables to nearby shore facilities and then beamed by FLTSATCOM satellite to a data-processing station in the United States. There a computer analyzes the information frequency by frequency, discriminating between sub noises and others. Not only does the process yield the type of submarine detected, but it often reveals the very name of the vessel in question.

Acutely aware of SOSUS's capabilities, Soviet submariners invented imaginative schemes to elude them. Some, hoping to reduce the noise they make below the threshold of SOSUS, have tried to creep past the arrays. Yet even at speeds as low as ten knots, spinning machinery inside the boat can emit tones in the register of a cello loud enough for the audience of hydrophones to overhear. Other subs have shut down their engines completely to drift past the sensors on silent ocean currents. And submarines looking to sneak by the array guarding the Strait of Gibraltar and another one stretching from the southernmost island in the Azores chain more than a thousand miles to the west frequently time their trips to coincide with the passage of large oceangoing surface ships whose noise might mask their own.

Despite such tactics, chances remain good that most submarines will be detected—if not heard by SOSUS, then revealed by satellite photographs to have left port or sensed by some other method that remains a closely held secret. During peacetime, only a few Soviet submarines are intercepted by American ones. "Their departure from base is a lot easier to monitor with satellites," states Pariseau. "We would send a boat to intercept and trail only if it could collect something unique—for example, what a new sub's acoustic signature is like, how deep it dives, and how fast it can go. These might be reasons to try to intercept."

Perhaps the *Lapon* had been given such an assignment. In any event—and however she had been directed to the Yankee-class boat—the attack sub had been on the boomer's tail for more than a month, an extraordinarily long time for one sub to track another without being detected, and a performance abetted by an uncommon predictability about the *Lapon*'s quarry.

Submariners the world over frequently perform chores and duties—clearing baffles, dumping garbage, receiving satellite communications, blowing sanitary tanks, and moving ballast water between the sub's trim tanks to preserve the delicate balance that keeps the bow of a submerged vessel from pointing too steeply upward or downward. The tasks are ordinary parts of submarine operations: They safeguard the health of the crew, preserve the stability and maneuverability of the boat, and keep it in touch with the home command. But they can also be noisy, and like any tasks that are repeated time after time, they can easily become routine.

American crews, believing that unpredictability enhances

stealth, have always taken pains to avoid this. For example, in the days before towed arrays were common, a submarine skipper, to make sure that no hostile submarine had sneaked up behind him, might instruct the officer of the watch to clear baffles three times during his six-hour tour of duty. It would then be the watch officer's responsibility to make sure that he did not schedule the task once every two hours. To make themselves unpredictable as to tactics as well as schedule, some submariners rolled dice to determine how they would clear baffles. "If the dice came up odd, you turned left," says Pariseau. "If they came up even, you turned right. And the number told you how many degrees you changed course before returning to your original heading. You'd do this just to make sure you didn't fall into the trap of becoming routine."

As the days went by, it became obvious that the crew of the Yankee had little regard for such considerations; the boomer performed the same tasks and maneuvers at regular intervals, day in and day out. After a while, members of the tracking party aboard the *Lapon* grew so familiar with their unseen opponent that they were able to do something that is extremely rare in this arena: They recognized individual officers of the deck from the way they handled the conn. The Navy remains tight-lipped about exactly how, but analysts speculate that one of the watch sections aboard the other sub may have been noticeably louder than the others, or that one officer consistently turned to the left when clearing baffles while the others went right. Or, the observers say, some officers might have given themselves away by the manner in which they brought the boat shallow to eject garbage or copy communications.

Able to forecast what the Yankee would do next and when, the *Lapon* succeeded in evading the boomer's efforts to detect her and tracked the missile boat throughout most of its Atlantic cruise—more than forty days in all. This is not to say that the Soviets never surprised the Americans, however: On more than one occasion when the boomer cleared baffles, it narrowly missed running into the *Lapon*, even though the attack sub had anticipated the maneuver and, following standard procedures, tried to move out of the Yankee's path. "You could hear him so close it sounded like the screw was going right through our hull," recalled a crew member. But even then, the Yankee did not detect the *Lapon*, which gave up shadowing the boomer only when it headed north through the Norwegian Sea, homeward bound at the end of its voyage.

Years after the *Lapon* adventure, the Soviets withdrew their elderly and noisy Yankees from the Atlantic and Pacific oceans and assigned them to patrols in European and Asian waters, where their missiles could not reach the American mainland. But the move only increased the challenge to the U.S. Navy's attack submarines. To replace the Yankees came a series of progressively larger ballistic-missile submarines capable of launching more and more dangerous nuclear weapons.

The first of these lethal new devices, the SS-N-8 deployed aboard the Delta I and Delta II subs, carried a pair of 800-kiloton warheads called independently targetable reentry vehicles because they could be steered individually toward separate aiming points. And subsequent Soviet boomers, including the 11,700-ton Delta IIIs, the 12,150-ton Delta IVs, and the latest 26,500-ton Typhoons—the largest submarines ever constructed—put to sea armed with nuclear-tipped missiles that could deliver even more warheads. The Typhoon's SS-N-20 can carry as many as nine reentry vehicles, each bearing a 100-kiloton warhead, and the Delta IV's SS-N-23 missile hauls ten such devices.

Most important, however, all of the new Soviet submarine-launched ballistic missiles can fly more than 5,000 miles—over three times the range of the Yankee's SS-N-6. The benefits that these truly intercontinental sea-launched ballistic missiles offered the Soviet Navy—and the difficulties they presented to American defenses—were obvious: The Deltas and Typhoons would no longer have to venture abroad to fire at targets in the Northern Hemisphere. Instead, they could hide in areas called bastions, some of which are located beneath the thick ice pack that covers 70 percent of the Arctic Ocean year round and extends southward in winter to the Barents Sea and the Sea of Okhotsk, waters that are no more than an easy day's sail from Murmansk, Vladivostok, or Petropavlovsk, the Soviet Union's three major submarine bases.

Under the ice, a submarine is immune to detection by snooping antisubmarine aircraft and surface ships. Planes can roam widely over the icecap but are poor sub hunters. Their primary sensor for localizing targets in the open ocean, the sonobuoy, is incapable of hearing sounds through a floor of thick ice, and no ice-penetrating versions of the device are yet operational. Instead, all the pilot of a plane or helicopter can do is sweep low over the icecap in search of a patch of open water that is not too far from a sub's suspected

The Soviet Typhoon-class ballistic-missile submarine *(left)* is the largest submersible ever built. Displacing 26,500 tons—as much as a World War II aircraft carrier—the 563-foot-long boats can carry twenty missiles, each having as many as nine nuclear warheads and a range of more than 5,000 miles.

location. If he finds such a pool and his aim is good, he might be able to drop a sonobuoy through it. However, basing a defense on such conjecture would be folly.

Vessels on the surface, unable to move through the ice, must confine their searches to the edge of the pack, an area known as the marginal ice zone, or MIZ. There the open ocean gives way to free-floating chunks of sea ice that become more numerous and less spread out the closer they are to the icecap. Waves slap constantly against the ice, putting up a persistent din up to eight times louder than the noise level of the open ocean—sufficient to drown out faint submarine sounds and thwart the effectiveness of the passive sonar arrays that surface ships tow. Moreover, many of the ice blocks are heavy enough to crush a hull or damage a propeller, and they often lie flush with the surface, making them hard to detect on radar and nearly impossible to spot with the naked eye. As a result, commanders of antisubmarine ships, none of which are built as icebreakers, typically give the ice pack wide berth.

Thus the only means available for ferreting out missile subs that hide in the Arctic is the attack submarine. It alone can leave the clamorous MIZ behind and gain entrance to the exceptional environment located farther in. Devoid of churning surface traffic and insulated from wave-inducing winds, which in the winter can blow with hurricane force, the water under the permanent icecap remains remarkably quiet. Though the ceiling above occasionally cracks open and grinds together loudly, background noise levels are usually some eight times lower than those in the open ocean—"quieter than dead quiet," say submariners. This silence unmasks sounds that are inaudible elsewhere and that can be heard over extraordinarily long distances, perhaps as great as 1,500 miles.

In time of war, Soviet boomers could easily have entered their under-ice bastions before American attack subs not already on patrol in the area could arrive. For instance, to reach the Beaufort Sea, north of Alaska—a body of water that would serve as a likely entryway to the Arctic for American subs—a Los Angeles-class attack ship departing Norfolk has to travel along the eastern seaboard past Newfoundland, then through the Davis Strait, along Greenland's west coast, and through the Barrow Strait, which skirts the southern edge of the Queen Elizabeth Islands—a two-week voyage. A boat based at America's other major submarine base, in San Francisco, has to cross the North Pacific and then feel its way through the shallow Bering Strait, the only passageway from the Pacific to the North Pole—a three-week sail.

But mitigating circumstances abound that would have given U.S. subs a good chance of arriving in time. In all likelihood, for example, they would have set out for the Arctic at some early sign of an irretrievably deteriorating political climate, not waiting for all the Soviet missile fleet to put to sea. Furthermore, sub-launched ballistic missiles on either side would have been fired, not early in a nuclear war, but sometime after the first exchange.

Slipping under the ice to begin searching for the boomers, the attack sub force could expect to encounter a defensive screen of the Soviets' most capable nuclear-powered predators—the Akulas and Victors—and a handful of diesel-electric boats, including the 3,000-ton Kilo-class patrol sub, which, when running on batteries, can be quieter than most nuclear ships.

Sonar operators could distinguish such vessels—which would be prowling about and making power-train noises, from the boomers, which would be stationary—and try to avoid them while seeking out the primary target. In this task, formidable as it seems, a number of factors would side with the American boats. First, waters under the icecap offer one of the best sonar environments on earth. Second, previous peacetime experience with tracking down Soviet boomers under the ice would have taught U.S. forces where under the ice their adversaries preferred to station themselves.

Considering the demonstrated superiority of American Los Angeles-class submarines in stealthiness, in the skills of their crews, in the sensitivity of their sonar to give them the first shot in combat, and in the nearly one-shot-one-kill trustworthiness of their torpedoes, U.S. Navy planners expected that American attack subs sent to sink the boomers would pull off the mission, even considering that some of the attacking force would be lost to the Soviets' defensive screen.

In order to elude enemy pickets, find a boomer, and approach within torpedo range, American skippers—more than the Soviets, who would already have taken their positions—would confront the considerable difficulties and risks of piloting a submarine under the icecap. Pushed by currents below and winds above, the pack moves constantly, sometimes traveling as far as ten miles in one day. As the ice floes drift, they inevitably pull apart, creating occasional long cracks called leads. The rifts briefly expose the ocean's surface until it is quickly covered over again by brittle new ice. When the floes are driven back together again, they crush the thin sheet and compress it into gigantic, solid ridges that can jut dozens of feet into the air and often reach downward more than 150 feet into the sea.

Only by making near-constant use of a special forward-looking active sonar can a Los Angeles-class boat safely transit ice-covered waters. The device, which is built into the front of the ship's sail, enables seamen to detect obstacles in their path by throwing a narrow beam of sound directly ahead of the submarine. Emitted at a high frequency, the sound ray has a range of only a few hundred feet and would rarely betray the ship's presence. Shaped like a cone, fifteen degrees wide at its narrowest, the beam expands steadily as it propagates, so that the farther it travels, the closer it comes to the surface. If the beam strikes a massive ridge thrust down by colliding ice floes, an echo returns to the sub and appears on a display, whereupon a crewman notes the distance to the contact.

Depending on the depth of the sub as it approaches such obstacles, those that dip only a short way into the water soon disappear from the display, as the beam no longer strikes them but passes underneath. Ice that plunges deeper remains on the screen longer, returning an echo even after the sub has closed to within a moderate distance and the beam hitting the obstacles is considerably narrower. As long as the contacts vanish from the display before the sub comes any nearer, however, the ship will still be able to pass harmlessly underneath. If they do not fade by the time the boat has approached within 300 yards or so, the crew must maneuver their vessel either around or below the looming obstacles.

To help pass under an ice keel without scraping the bottom, a sub is equipped with two additional high-frequency sonars, a Fathometer in the keel that reveals how far down the bottom lies and a similar device mounted at the top of the sail. Called an ice profiler, it reports on the height of the ice ceiling overhead. Before the invention of forward-looking sonar, these instruments were the only aids available to help a sub skipper pass unscathed under the icecap.

Such was the situation in the summer of 1957 when Captain Bill Anderson, commander of the USS *Nautilus*, made the first of his two attempts to pilot his submarine under the North Pole. Anderson intended to transit the Canadian Basin, one of four valleys that plummet as far as 14,000 feet beneath the surface of the five-million-square-mile Arctic Ocean. But to reach the deep water from

Hawaii, he first had to cross the Bering Strait and the Chukchi Sea, waters that, because of the broad, flat shelf that skirts the Asian and North American continents, average a depth of only about 110 feet.

Ice keels presented a daunting test of seamanship for the crew of America's first nuclear submarine. Using only a Fathometer and an ice profiler to feel his way, Anderson could detect ice keels only when he was nearly upon them. Even so, he once guided the *Nautilus* beneath an 80-foot-deep ridge in just 142 feet of water. The boat inched by with only six feet of clearance between its belly and the seabed, and another half-dozen between the top of the sail and the jagged ice. Unwilling to subject his ship to such risks, he aborted the mission and returned to Pearl Harbor, where the *Nautilus* was fitted with additional and more sensitive sonars. With this improvement and others, he made a second, successful assault on the pole in August 1958.

The upward-looking sonar mounted at the top of the sail can also distinguish between thick ice, which produces a strong, clear echo, and thin ice or open water, both of which send back weak echoes or none at all. (A tiny video camera mounted in the observation periscope supplements the sonar. In summer, when the midnight sun illuminates the Arctic, bright images on TV monitors in the skipper's quarters and control room indicate thin ice or open water.)

Leads and irregularly shaped expanses of thin ice called polynyas are helpful to a submarine, permitting it to extend a radio mast from periscope depth, for example, get a celestial navigation fix on the stars, or perform some other task—and even to surface in the event of an emergency. "You always keep track of polynyas so that you know where the closest polynya is," says Beers, who has broken through the icecap many times, "so that you can go back there." Often the polynya has closed, sending the sub in search for another.

To measure a lead or a polynya and determine whether it is large enough for a sub to surface, the boat typically runs at a constant speed back and forth below it. During each pass, crewmen carefully note the ship's course and speed, as well as the times at which the returns from the under-ice sonar weaken and then grow stronger again—and vice versa. Plotting these points to scale on a chart shows the size and shape of the opening in the ice.

If the opening is judged big enough, the submarine will come around one last time and stop directly beneath the polynya. Then, while the wind and current push and pull on the ice overhead, changing the shape of the opening, the crew concentrates on keeping the ship in position to rise through it. The cost of shoddy seamanship here can be high: If the boat moves laterally, forward, or backward while going up, the force of the impact with the ice might not be transmitted vertically through the sail, which is heavily reinforced for the job, but at an angle. This could damage some of the submarine's delicate topside sensors or knock the masts in the sail out of alignment.

Inside the control room, special attention is also devoted to the ship's trim. If the planesman fails to keep the vessel's nose at exactly the same depth as its tail while the boat ascends, the sub will run the risk of rapping the ice not only with its sail, but with the comparatively fragile stern or bow as well.

Diving planes are turned vertical to slip through the ice, and the submarine is ready to surface. To make sure the boat does not ascend faster than about ten feet per minute—a speed chosen to avoid damage to the top of the sail when it comes into contact with the ice ceiling—the crew pumps water out of the sub's trim tanks. "If you used your main ballast tanks, you'd come up too fast," Captain Gustavson says.

"The underside of the ice canopy is soft," he continues. "If you have ice that is, say, four feet thick, there might be an additional foot of mushy, slushy ice-water interface. It's actually a pretty stable evolution to come up underneath the ice and poke your sail into it. When you hit the ice, you start punching through it until you stop." After that, a few short blows with the main ballast tanks a couple of seconds apart are usually enough to fracture the ice completely and ram the sail up into the chill polar air.

Searching for boomers under the ice in wartime, an attack submarine would be disinclined to surface. The quarry, after all, would be hiding underwater, almost certainly nestled against the ice canopy and between jagged ice keels that project deeper than the boomer's keel. Hunting might be best near leads or polynyas large enough to accommodate a missile sub when it surfaces to launch its weapons.

This tactic of concealment, known as ice-picking, offers a considerable impediment to detection, yet the missile boat is not whol-

Crew members of the USS *Billfish* prepare to reboard after an Arctic rendezvous with another submarine *(not shown)* to check the accuracy of the boat's inertial navigation system. To surface, the submarine punched through some four feet of ice, assisted by a steel cap on top of the sail and by diving planes that turn vertical to prevent the weight of ice from breaking them.

ly indiscernible. Inside, a turbine generator continues to make electricity, the desalinization plant goes on producing fresh water, and the oxygen generators and carbon dioxide scrubbers still clean and replenish the air supply. Moreover, the members of the crew carry on as usual: They do routine housekeeping, run pumps, and use other internal machinery—all of which make noise, however faint or occasional, that a hunter of the deep can use to zero in on its prey.

"You figure it out the same way you do your normal tactics," explains Gustavson. "You listen passively and maneuver to solve for range. And as you drive back and forth, all of your bearing lines start to lay down at the same point." The difference is that the ice-picking process takes more time, minutes or hours that may be in short supply in combat.

The longer a tracking party spends taking such bearings, the more precise becomes its estimate of the target's location. Their objective is familiar to all submariners, regardless of the arena: to achieve a firing position that, if need be, would give their torpedoes the best possible chance of homing on the target and leave the enemy with little or no time to escape or counterattack.

For fighting other subs, the Navy's Mark 48 torpedo carries a 600-pound shaped-charge warhead to burn a hole through an enemy sub's pressure hull and can search for its prey using both active and passive sonar. A Soviet boomer ice-picked in a stand of ice keels would be a particularly challenging target. To keep the Mark 48 from inadvertently colliding with one of them, an attack-sub skipper would typically preprogram the weapon to run at a depth far below the deepest of these projections, and to start searching with passive sonar only after it has covered most of the distance to the target. "We tell it where to run," says Beers. "We take the sonar conditions on the ship and optimize the torpedo's run so that it has the strongest possibility of receiving the signal from the target."

Turning slightly nose up, the torpedo would listen for the enemy sub. As the torpedo neared the target, the sounds that betrayed it to sonar operators aboard the attack sub would become easier for the weapon's passive sonar to pick out, leading the torpedo in the direction of the boomer. Arriving almost directly below the missile sub, the Mark 48 would nose nearly straight up toward the target.

Until now, the torpedo's active sonar would have been silent. "A sub stopped underneath the ice," says Beers, "looks the same as a piece of ice; a torpedo has no way of discriminating between the

A crewman on the USS *Oklahoma City* checks screws at the breech of a torpedo tube before the Mark 48 behind him is loaded. Teflon applied to the rail-like structures inside the tube and to the torpedo lubricates the passage of the weapon through the tube.

two. They're both heavy objects that will give a suitable return, so the torpedo might home on either one—the ice or the sub." At a preprogrammed depth—with the torpedo so close to the boomer that its active sonar field of view could see nothing of nearby ice keels, but only the target—the weapon would begin to ping. "You try to set the torpedo to turn on very close to the target," explains Beers, "so the only thing the torpedo sees is another sub."

Then, as sonar operators back on the sub listen in, the weapon would close on the target, pinging rhythmically. At first, the echoes would come fractions of a second after the pings, but as the range closed, the two sounds would ring through the water like a high-pitched drum roll—a beat that would end not with a clash of cymbals, but with the terrific roar of a hull-bursting explosion.

For years, the image of Soviet boomers letting off a salvo of long-range ballistic missiles was a recurring feature of Cold War nightmares. Fortunately, the chances of such bad dreams becoming reality today are slim. Yet as long as strategic subs such as the Russian Deltas and Typhoons (as well as the American Ohio-class ships) remain in service, their skippers are certain to go on observing the rituals they and their comrades have rehearsed so many times during the past three decades—and to hold the ardent attention of the other side's attack submarines. ★

Arrows for
the Quiver

Simple to say, hard to do, submarines prowled the oceans in World War II to sink ships. Fire control consisted of some lines on a periscope lens, an optical range finder, a stopwatch, and a quick prayer that the torpedo—a non-homing "tin fish"—would run true. With much skill and a little luck, the captain would be rewarded with a kill.

Since then, new technologies have broadened the attack sub's variety of victims to include not only surface ships but other submarines and even targets ashore. Fire control today means computers that analyze and integrate information from as many as five sonar systems.

Submarine-delivered weapons have also improved dra-

Detonated by a proximity fuse, a British Tigerfish torpedo explodes below the keel of a target ship, lifting the vessel near the center with force enough to break its back. Such damage is invariably fatal.

matically. As explained on the following pages for weapons carried by American Los Angeles-class subs, range has been extended from a few thousand yards, in the case of torpedoes, to many miles—and to hundreds of miles for missiles launched at enemy ships and land facilities. Dumb tin fish that have to be aimed at the quarry are a thing of the past; nowadays, torpedoes home on the enemy with sonar, hanging on more tenaciously than hounds trailing a fox. There are mines that fire torpedoes at enemy subs, and the phenomenal accuracy of the Tomahawk cruise missile means almost certain catastrophe for anything, afloat or ashore, that it sets out to destroy.

A Submariner's Most Satisfying Punch

The torpedo room of a Los Angeles-class attack sub is a blue-collar environment of heavy equipment and straining muscles. In seeming chaos, pipes, wires, and hoses run in bundles along the ceiling and walls. Torpedoes weighing nearly two tons apiece are strapped down in three two-tiered, motorized racks, which also store torpedo-shaped mines, decoys, and missiles.

When the order is given to load a torpedo, the room comes alive. One of the racks moves a torpedo to the breech of a torpedo tube, then pivots to align the weapon for sliding into the tube on rollers built into the rack. As the breech is closed, electrical contacts in the door link the torpedo's guidance computer with the sub's fire-control system, which supplies information on the target's position, course, and speed.

As shown at right, loading and firing a torpedo requires an intricate system of pipes, valves, and pistons that, upon the captain's command, flush the weapon from the tube by means of water pressure. Outside, a chemical motor starts, and the torpedo heads for the target.

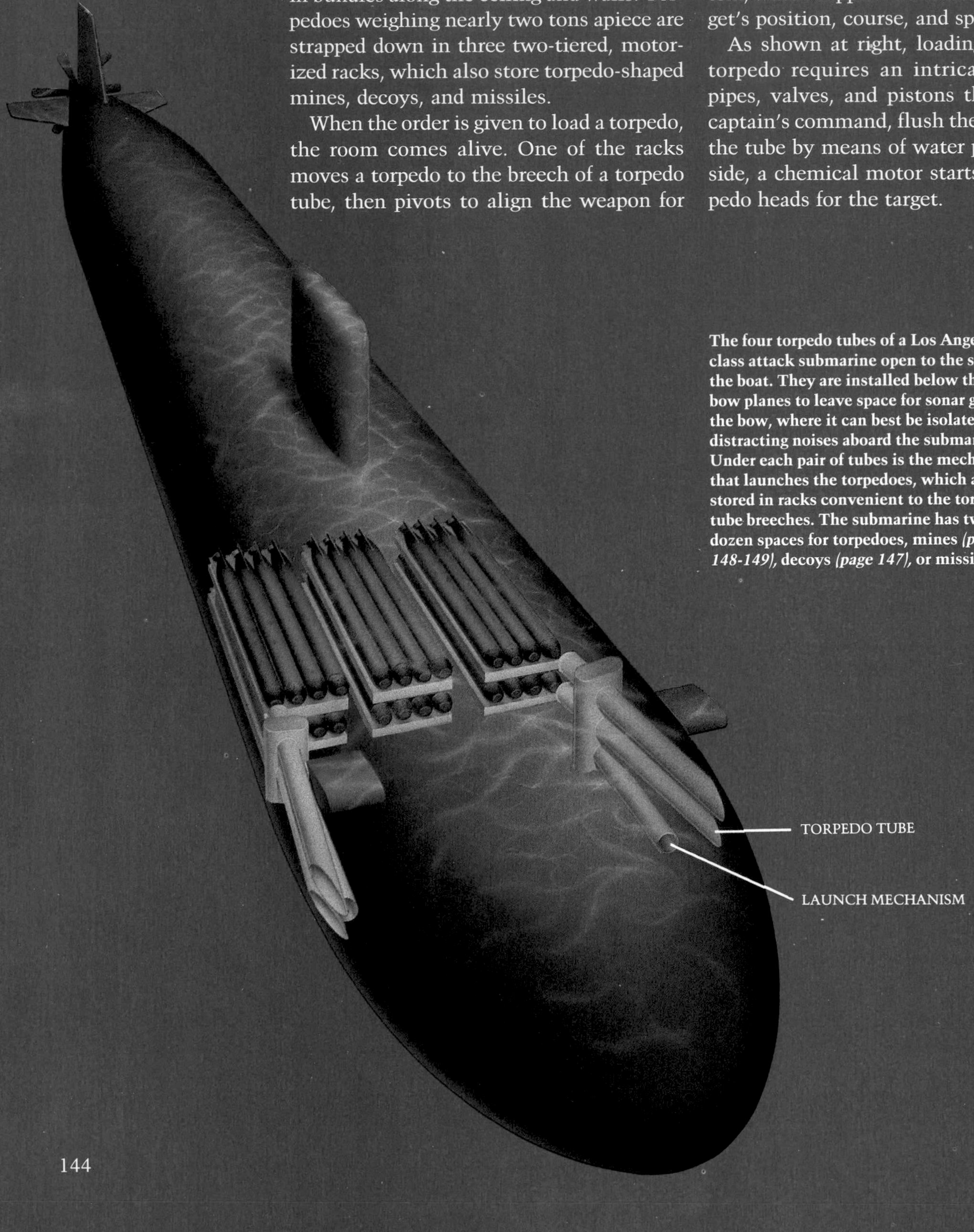

The four torpedo tubes of a Los Angeles-class attack submarine open to the side of the boat. They are installed below the bow planes to leave space for sonar gear in the bow, where it can best be isolated from distracting noises aboard the submarine. Under each pair of tubes is the mechanism that launches the torpedoes, which are stored in racks convenient to the torpedo-tube breeches. The submarine has two dozen spaces for torpedoes, mines (pages 148-149), decoys (page 147), or missiles.

TORPEDO TUBE

LAUNCH MECHANISM

The Firing Sequence

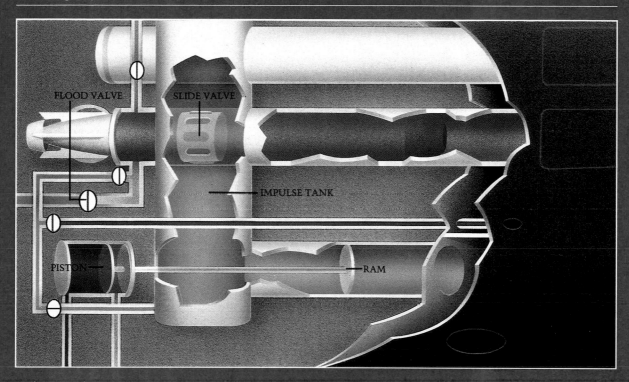

FLOOD VALVE   **SLIDE VALVE**

IMPULSE TANK

PISTON

RAM

Loading a torpedo. As the breech of a torpedo tube is opened for loading, the tube contains low-pressure air *(pink)*. High-pressure air *(red)* to the left of a piston holds a ram, linked to the piston by a rod, at the ready position inside a tube extending from the base of a water-filled container called the impulse tank. A slide valve and a flood valve prevent water from entering the tube as the torpedo is slipped inside and the breech closed.

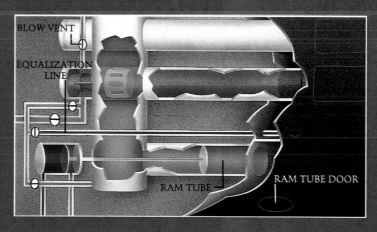

BLOW VENT

EQUALIZATION LINE

RAM TUBE   **RAM TUBE DOOR**

Flooding the tube. After the breech is sealed, the flood valve is opened and water under low pressure *(light blue)* flows into the tube, forcing air out a pipe called a blow vent. When the torpedo tube is full, the blow vent and flood valve are closed. Next, a valve in an equalization line and the ram-tube door are opened to equalize pressure in the system with the higher sea pressure outside *(dark blue)*.

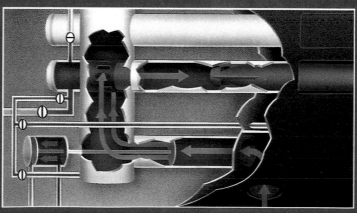

Firing the shot. To prepare for firing, all open valves are closed, then the slide valve and the outer door of the torpedo tube are opened. At the moment of launch, high-pressure air is piped to the right side of the piston, pushing it and the attached ram to the left. The motion of the ram forces water through the slide valve into the torpedo tube, flushing the torpedo.

A Smart Fish for Killing Subs

The Mark 48 ADCAP (added capability) torpedo is the most advanced such weapon in the U.S. arsenal. With a twenty-mile range and a top speed of fifty-five knots, it costs about $2 million. Intended primarily for use against submarines, the Mark 48 carries a high-explosive warhead. It guides itself but can be steered manually for short distances over a thin wire that unreels behind it.

To find the quarry, the Mark 48 ADCAP has a phased-array sonar. In its passive mode, explained below, the sonar can determine the direction of a sound emanating within a cone of about ninety degrees extending from the nose of the weapon. In the active mode, the sonar can ping a target with sound pulses anywhere in the cone and detect echoes from several miles away.

Such performance is a great improvement on sonar of older weapons, which could hear only what was directly ahead of them, so a torpedo had to snake through the water to seek a target. A phased-array sonar eliminates this snaking and thus accounts in large measure for the Mark 48's considerable range and swift arrival at the target.

PROPELLER SHROUD

FUEL

ENGINE

GUIDANCE COMPUTER

WARHEAD

STEERING CONTROL

SONAR ARRAY

Inside a Mark 48. From sonar in the nose to propellers at the tail, the Mark 48 is packed with gear for its mission. The weapon's fuel contains its own oxygen; when ignited, it produces heat that generates steam for an engine connected to the propellers.

Wide-Angle Hearing

In the illustrations of phased-array sonar at right, a sound wave front arrives at three hydrophones as two previous wave fronts travel in the form of electric impulses to three collectors called summers. The wiring leading to the left and right summers has delay units that slow the impulses to varying degrees. In the diagram at near right, the simultaneous arrival of the undelayed impulses at the center summer creates the loudest response, alerting the sonar that the sound source is dead ahead. At far right, sound waves from an off-center source arrive at the right summer in unison, indicating a target to one side of the sonar.

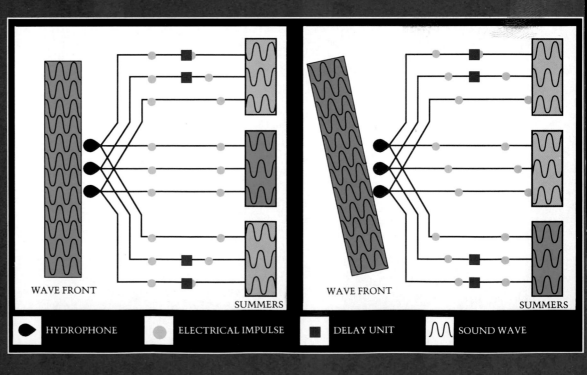

WAVE FRONT

SUMMERS

WAVE FRONT

SUMMERS

● HYDROPHONE   ● ELECTRICAL IMPULSE   ■ DELAY UNIT   〰 SOUND WAVE

A Relentless Pursuit

1 After detecting an enemy sub, a Los Angeles-class boat fires a Mark 48 ADCAP torpedo. It runs to starboard before turning toward the target so that, when heard by the enemy, it will not reveal the attacker's bearing.

2 As the Mark 48 runs at full speed to the predicted location of its target, the enemy sub easily detects the loud whine of the torpedo's engine and propellers. Deploying a decoy, the boat makes for a canyon of deep water, even as the torpedo slows to begin pinging for its prey.

3 Hearing the pinging, the decoy responds with sounds that mimic the sonar echo from a sub. The torpedo would veer toward the noise except for a sonar operator aboard the Los Angeles. Suspicious of the echo's loudness, he takes control of the Mark 48, turns off the sonar, and steers the weapon around the decoy.

4 As the sonar operator reactivates the Mark 48's sonar and frees the weapon to resume the hunt, the torpedo detects the enemy sub just before it slips over the lip of the canyon and out of sonar range. The torpedo accelerates to the top speed at which it can still discern echoes from the sub over its own din.

5 The torpedo continually confirms its target's presence by pinging more and more rapidly as it approaches. Just before impact, ping and echo seem to merge into one sound that means disaster for the enemy.

Weapons of Patience

Mines are best sown secretly where only the enemy is likely to sail. A prime candidate for the job is the submarine, which can sneak in underwater and plant mines even at harbor entrances. Submarine-delivered mines come in two basic varieties, both expelled through the boat's torpedo tubes. One, called the submarine-launched mobile mine (SLMM), is for use against ships; the other sinks submarines.

Antisubmarine mines like the CAPTOR (for encapsulated torpedo) go after their quarry much as submarines do: They listen intently for a boat approaching, then fire at it with a sonar-homing torpedo.

These mines are difficult to sweep because they can distinguish submarine-like sounds from other sea and ship noises. So the usual method for dealing with mines—a helicopter-towed raft that broadcasts ship noises to trick the devices—has little success, since the mine capsule's active sonar can distinguish between a raft and a sub's

hull. The most effective way to neutralize CAPTOR-style mines is a device similar to the decoy described on page 147. But decoys are costly, the more so because they are destroyed if they find mines.

Antiship mines, besides listening with hydrophones, often use two other means of detecting approaching vessels: magnetism and pressure. Ships become magnetized as a result of being enveloped by the earth's magnetic field. Demagnetizing a ship from time to time makes it less likely to trigger such a mine, but the mere presence of a vessel's steel hull, demagnetized or not, may be enough to set it off.

Pressure sensors detect the subtle rise and fall in water pressure caused by the passage of a ship. This trigger is impossible to counter. Only a passing ship can cause the progression that will detonate a pressure-influenced mine, and there is no way to prevent a ship from producing the pattern.

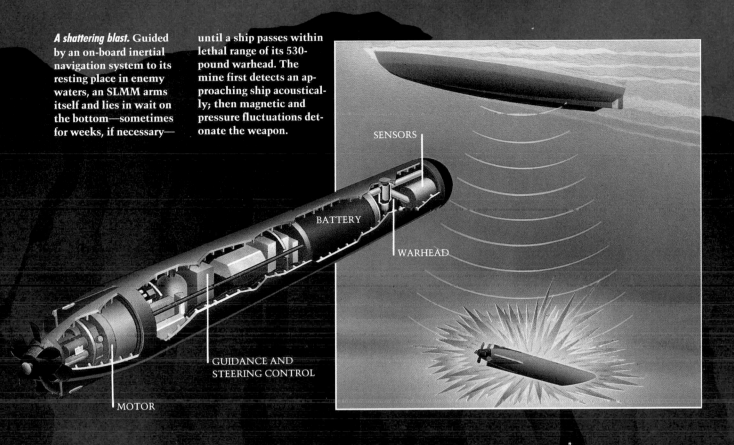

A shattering blast. Guided by an on-board inertial navigation system to its resting place in enemy waters, an SLMM arms itself and lies in wait on the bottom—sometimes for weeks, if necessary— until a ship passes within lethal range of its 530-pound warhead. The mine first detects an approaching ship acoustically; then magnetic and pressure fluctuations detonate the weapon.

SENSORS

BATTERY

WARHEAD

GUIDANCE AND STEERING CONTROL

MOTOR

A direct hit. Heading for the bottom as soon as it exits the torpedo tube, the CAPTOR descends to a predetermined depth, and the mine's weighted bottom detaches from the buoyant capsule, anchoring it to the seabed. The mine then switches on the hydrophones of its passive sonar. When the presence of a sub is confirmed, a lightweight Mark 46 torpedo swims out of the capsule and attacks the boat, homing with its own active sonar.

TORPEDO

ANCHOR

HYDROPHONE

BATTERIES

Over the Horizon and Beyond

Tomahawk cruise missiles provide submarines of the Los Angeles class with a long reach. A land-attack version of the jet-powered, subsonic weapon can strike at targets up to 800 miles away. Largely because the antiship model must follow a serpentine flight path to find the target, its range is only 300 miles, about fifteen times that of a Mark 48 torpedo.

Sealed inside twelve vertical tubes designed for submerged launching (opposite), the two types differ chiefly in their guidance systems. Each has a gyroscopic inertial navigation system (INS). In addition, the land-attack Tomahawk has two other guidance systems. Terrain contour matching (TERCOM) uses digital relief maps. The digital scene matching area correlator (DSMAC) guides the missile with digitized photographs.

Because surface vessels are often under way when attacked—and because TERCOM and DSMAC can find only stationary objects—the antiship Tomahawk relies on radar to pinpoint its target. Both models are shown in operation on pages 152-153.

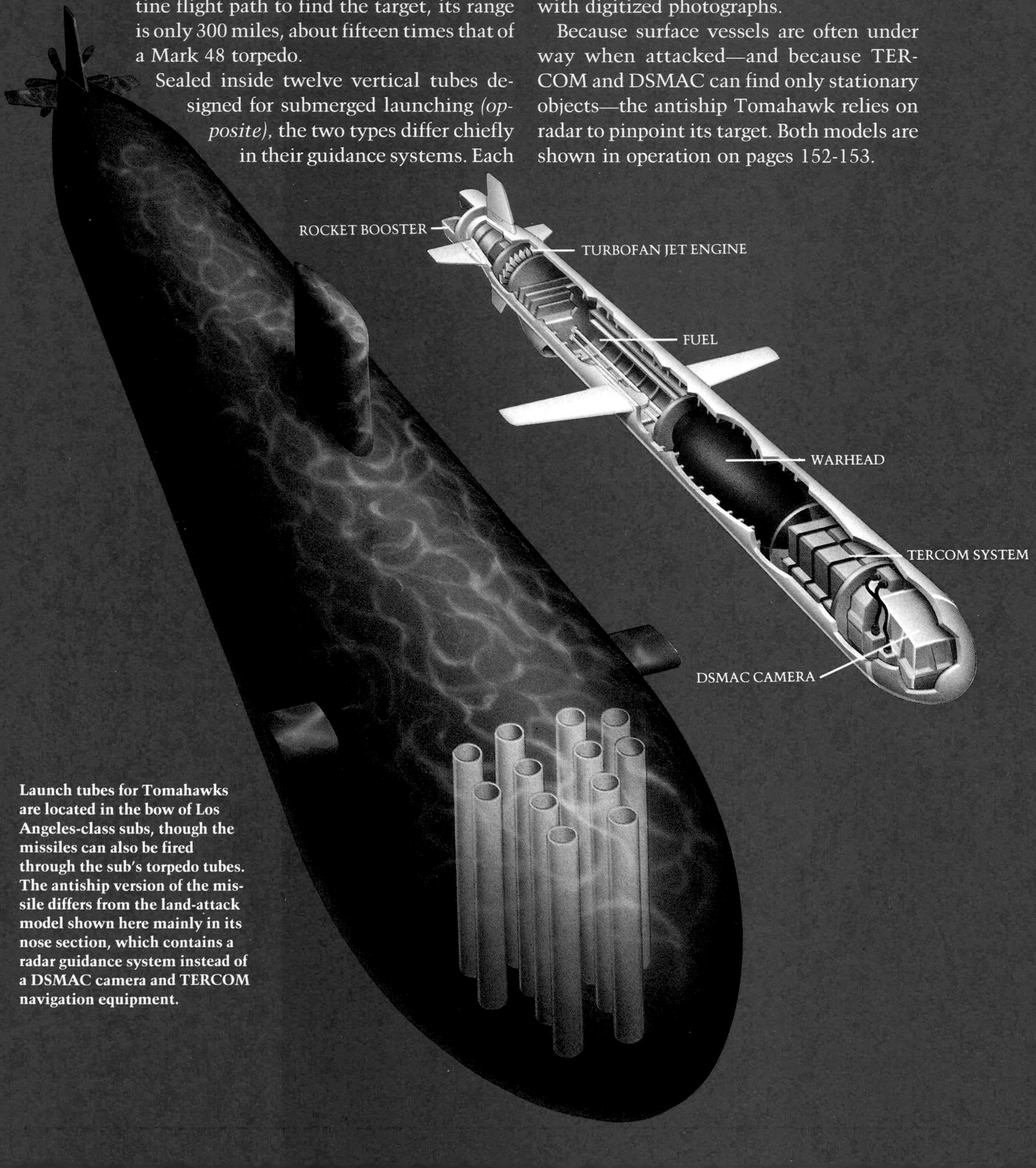

ROCKET BOOSTER

TURBOFAN JET ENGINE

FUEL

WARHEAD

TERCOM SYSTEM

DSMAC CAMERA

Launch tubes for Tomahawks are located in the bow of Los Angeles-class subs, though the missiles can also be fired through the sub's torpedo tubes. The antiship version of the missile differs from the land-attack model shown here mainly in its nose section, which contains a radar guidance system instead of a DSMAC camera and TERCOM navigation equipment.

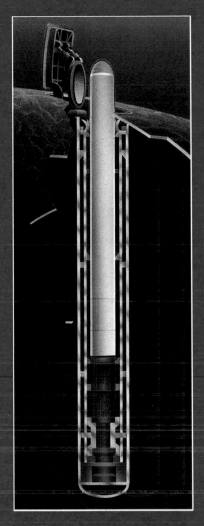

Out of the tube. When the hatch of a launch tube is opened at the firing depth of sixty feet, only a thin plastic membrane—prevented from collapsing by air pressure inside the capsule at about thirty pounds per square inch—separates the missile from the sea. Upon a command to launch, an explosive charge at the base of the capsule thrusts the missile upward, through the membrane, and away from the submarine.

Through the water. When the missile is about twenty-five feet above the sub, its booster rocket ignites and propels it to the surface. As soon as the missile pops into the air, tail fins extend to steady its upward flight.

Taking flight. At an altitude of about 1,000 feet, the rocket booster burns out and drops away. Stubby wings spread from slots in the missile fuselage, an air scoop deploys, and the turbofan engine spools up. Within seconds, the Tomahawk transitions to level flight, en route to its target.

Seeking Out a Ship

To sink a ship with a Tomahawk, a submarine commander must be informed by radio of its approximate location, course, and speed. The communication also contains a rundown of other ships in the vicinity. Coordinates of both the target and ships that the missile should avoid are quickly passed by computer to a computer governing the missile's inertial navigation system, and the commander gives the order to fire.

1 Shortly after the Tomahawk descends to radar-evading altitude, it nears an enemy trawler. Programmed in advance to avoid this ship and others, any of which could warn the target of the missile's approach, the Tomahawk's INS computer steers around the threat.

2 The Tomahawk climbs a couple of hundred feet for a better vantage point and activates its target-seeking radar. Then, the missile begins to fly a snaking search pattern programmed before launch to take in an area that is certain to contain the target, even if it changes course or speed.

3 Upon sighting the target, the Tomahawk updates its INS with the new position, then dives to sea-skimming altitude to evade enemy radar, which has probably been tracking the missile. To confuse enemy defenses, the Tomahawk flies a roundabout route to the target, guided by the INS computer.

4 A few hundred yards from the target, the Tomahawk pops up for a last look with radar. Then it dives for the ship, punching through the lightly armored deck and exploding deep inside.

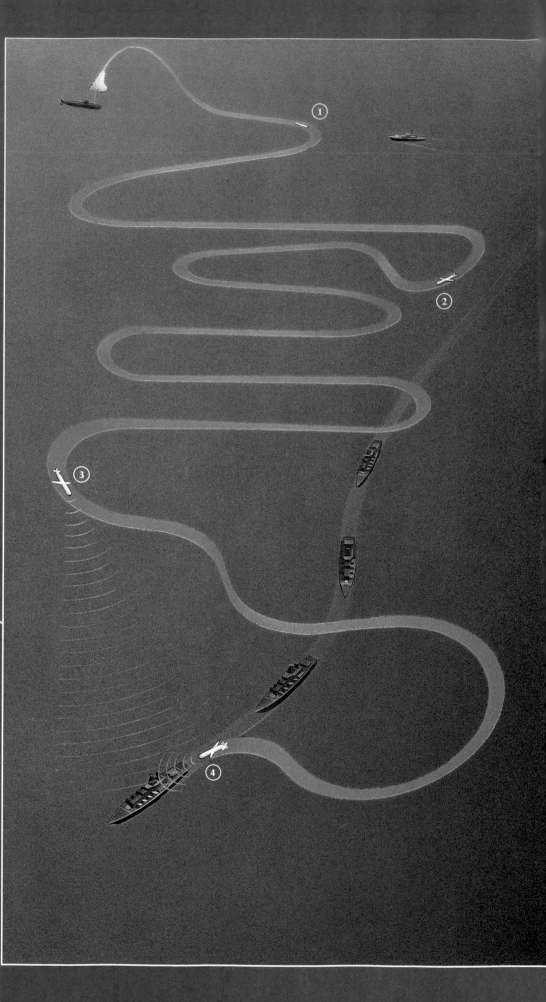

Finding a Target Ashore

Unlike the antiship version of the Tomahawk, the land-attack model requires a stationary target, such as a power plant or a headquarters building. To find it, the missile's INS uses two aids: several digital terrain maps for the TERCOM system, and one or more images for DSMAC.

1 Guided by INS to the first TERCOM map, the Tomahawk computer compares readings from on-board altimeters to numbers stored in memory representing the average height of terrain in small squares of the map. This permits TERCOM to judge how much the missile has strayed from its planned course across the first map. The system compensates in two ways: It adjusts the course as the missile turns toward the second map, but more important, it calculates and compensates for subtle inaccuracies unavoidably built into any INS.

2 Adjusting the accuracy of the INS permits the second TERCOM map to be smaller than the first and also to contain smaller squares. The finer detail yields more accurate position fixes and more precise compensation for INS error. Subsequent maps are smaller yet, with even finer detail, refining missile accuracy to a hundred feet or less.

3 Within a few miles of the target, the Tomahawk switches to DSMAC. Using a video camera triggered by the INS, the system takes two-tone images of the ground below *(inset)* and compares them with images stored in its memory. Adjusting the INS and the flight path as needed, DSMAC can achieve accuracy of as little as ten feet.

4 Knowing the precise position of both target and missile, the INS flies the Tomahawk toward the mark along a preplanned path. So accurate is the Tomahawk that the bull's-eye can be a specific room in a building.

Acknowledgments

The editors of Time-Life Books wish to thank the following for their assistance in the preparation of this volume: Guy Aceto, *Air Force Magazine*, Arlington, Va.; Dave Baker, Pentagon, Washington, D.C.; Bob Bannister, Kollmorgen Systems, Northampton, Mass.; Captain Charles J. Beers, Washington, D.C.; Gerald Behm, Indian Head, Md.; Susan Boyd, Washington, D.C.; William P. Calvani, Nautilus Memorial and Submarine Force Library and Museum, Groton, Conn.; Lt. Deborah Carson, Pentagon; Lt. Comdr. David Clites, Washington, D.C.; Lt. Nick Connolly, Washington, D.C.; Dorothy Cross, Pentagon; Linda Cullen, Naval Institute Press, Annapolis, Md.; Conrad Curry, Naval Recruiting Command, Arlington, Va.; Lorna Dodt, Pentagon; Susan Eddy, Kollmorgen Systems, Northampton, Mass.; Russ Egner, Pentagon; Dr. Jerome Feldman, David Taylor Research Center, Bethesda, Md.; Susan Fili, Washington, D.C.; Lt. Frank Flaherty, Norfolk, Va.; Capt. George Graveson, Annandale, Va.; Comdr. Charles Griffiths, Washington, D.C.; Capt. Fred P. Gustavson, Washington, D.C.; WO Ron Henry, Norfolk, Va.; Lt. Thomas Heron, Norfolk, Va.; Chief Mike Hoffler, Norfolk, Va.; Lt. Comdr. Pat Hopfinger, USS *Oklahoma City*, New York, N.Y.; Hugh Howard, Pentagon; Lt. (jg) Jeffrey H. Hutchison, Norfolk, Va.; Yogi Kaufman, Yogi Inc., Potomac, Md.; Lt. (jg) Taylor Kiland, Pentagon; Walt Lang, Pentagon; Spiro Lekoudis, Arlington, Va.; Capt. Warren Lipscomb, Norfolk, Va.; Robert McCaig, IBM, Manassas, Va.; Adm. Paul McCarthy, Alexandria, Va.; Lt. Comdr. George MacEwen, Washington, D.C.; Patty Maddocks, Naval Institute Press, Annapolis, Md.; Capt. William Manthorpe, Pentagon; Irene Miner, Pentagon; Lyle Minter, Pentagon; Lt. Comdr. David Morris, Norfolk, Va.; Don Nolan, Naval Surface Warfare Center, White Oak, Md.; John F. O'Connell, Kapos Associates, Arlington, Va.; Diane Palermo, Naval Surface Warfare Center, White Oak, Md.; Chief David Pearson, Norfolk, Va.; TMC Drew Pearson, Norfolk, Va.; Lt. Comdr. Robert Perry, Washington, D.C.; Comdr. Phil Polefrone, USS *Oklahoma City*, New York, N.Y.; Chuck Porter, Anacostia Naval Station, Washington, D.C.; William Rosenmund, Pentagon; Neil Ruenzel, General Dynamics, Groton, Conn.; William Ruhe, McLean, Va.; Obaid Sadiq, Naval Attache-Pakistan; Sam Salem, Loral Defense Systems, Akron, Ohio; Dorothy Sappington, Naval Institute Press, Annapolis, Md.; M. R. Sharma, Alexandria, Va.; Lt. Greg Smith, Pentagon; Gary Stiegerwald, Naval Underwater Systems Center, Newport, R.I.; Mary Beth Straight, Naval Institute Press, Annapolis, Md.; Comdr. Lyal Stryker, Association of Minemen, North Charleston, S.C.; Gary Stubblefield, Applied Maritime Technology, Bonita, Calif.; Mabel Thomas, Pentagon; Pat Toombs, Pentagon; Capt. John Vick, Annandale, Va.; Bob Waller, Anacostia Naval Station, Washington, D.C.; Lt. Bo Williams, Norfolk, Va.; Kevin Young, IBM, Manassas, Va.

Bibliography

BOOKS

Anderson, William R., and Clay Blair, Jr., *Nautilus 90 North*. New York: Harper & Row, 1959.

Beach, Edward L., *Around the World Submerged*. New York: Holt, Rinehart and Winston, 1962.

Bishop, Chris, ed., *The Encyclopedia of World Sea Power*. New York: Crescent Books, 1988.

Bowditch, Nathaniel, *American Practical Navigator*. Brookmont, Md.: Defense Mapping Agency Hydrographic/Topographic Center, 1984.

Breemer, Jan, *Soviet Submarines*. Surrey, England: Jane's Information Group, 1989.

Brown, David, *The Royal Navy and the Falklands War*. London: Leo Cooper, 1987.

Burrows, William E., *Deep Black*. New York: Random House, 1986.

Collier's Encyclopedia. Vol. 21. New York: Macmillan Educational Company, 1986.

Compton-Hall, Richard:
Submarine versus Submarine. New York: Orion Books, 1988.
Submarine Warfare: Monsters & Midgets. Poole, England: Blandford Press, 1985.

Cox, Albert W., *Sonar and Underwater Sound*. Lexington, Mass.: Lexington Books, 1974.

Duncan, Francis, *Rickover and the Nuclear Navy*. Annapolis, Md.: Naval Institute Press, 1990.

English, Adrian, and Anthony Watts, *Battle for the Falklands (2) Naval Forces* (Men-at-Arms series).
London: Osprey Publishing, 1982.

Freedman, Lawrence, and Virginia Gamba-Stonehouse, *Signals of War*. Princeton, N.J.: Princeton University Press, 1991.

Frieden, David R., ed., *Principles of Naval Weapons Systems*. Annapolis, Md.: Naval Institute Press, 1985.

Friedman, Norman:
The Naval Institute Guide to World Naval Weapons Systems. Annapolis, Md.: Naval Institute Press, 1989.
Submarine Design and Development. Annapolis, Md.: Naval Institute Press, 1984.
U.S. Naval Weapons. Annapolis, Md.: Naval Institute Press, 1989.

Friedman, Richard S., et al., *Advanced Technology Warfare*. New York: Harmony Books, 1985.

Gates, P. J., and N. M. Lynn, *Ships, Submarines and the Sea*. London: Brassey's, 1990.

Hartmann, Gregory K., and Scott C. Truver, *Weapons That Wait*. Annapolis, Md.: Naval Institute Press, 1991.

Hastings, Max, and Simon Jenkins, *The Battle for the Falklands*. New York: W. W. Norton, 1983.

Hill, J. R., *Anti-Submarine Warfare*. Annapolis, Md.: Naval Institute Press, 1989.

Hobbs, Richard R., *Marine Navigation*. Annapolis, Md.: Naval Institute Press, 1990.

Jane's Fighting Ships. Ed. by Richard Sharpe. Surrey,

England: Jane's Information Group, 1990.

Jane's Underwater Warfare Systems. Ed. by Bernard Blake. Surrey, England: Jane's Information Group, 1990.

Kaufman, Steve, and Yogi Kaufman, *Silent Chase.* Charlottesville, Va.: Thomasson-Grant, 1989.

Keegan, John, *The Price of Admiralty.* London: Viking Penguin, 1988.

Lehman, John F., Jr., *Command of the Seas.* New York: Charles Scribner's Sons, 1988.

Leitenberg, Milton, *Soviet Submarine Operations in Swedish Waters, 1980-1986.* New York: Praeger, 1987.

McGraw-Hill Encyclopedia of Science and Technology. Vol. 17. New York: McGraw-Hill, 1987.

Maroon, Fred J., and Edward L. Beach, *Keepers of the Sea.* Annapolis, Md.: Naval Institute Press, 1983.

Middlebrook, Martin:
The Fight for the "Malvinas." London: Viking, 1989.
Operation Corporate. London: Viking, 1985.
Task Force: The Falklands War, 1982. London: Penguin Books, 1987.

Miller, David, *Modern Submarines.* New York: Prentice Hall, 1989.

Miller, David, and John Jordan, *Modern Submarine Warfare.* London: Salamander Books, 1987.

Miller, David, and Chris Miller, *Modern Naval Combat.* New York: Crescent Books, 1986.

Moore, John E., and Richard Compton-Hall, *Submarine Warfare: Today and Tomorrow.* Bethesda, Md.: Adler & Adler, 1987.

Moro, Rubén O., *The History of the South Atlantic Conflict.* New York: Praeger, 1989.

Oberg, James E., *Uncovering Soviet Disasters.* New York: Random House, 1988.

Perkins, Roger, *Operation Paraquat.* London: Picton Publishing, 1986.

Perrett, Bryan, *Weapons of the Falklands Conflict.* Poole, England: Blandford Press, 1982.

Polmar, Norman:
Atomic Submarines. Princeton, N.J.: D. Van Nostrand Company, 1963.
The Ships and Aircraft of the U.S. Fleet. Annapolis, Md.: Naval Institute Press, 1987.

Polmar, Norman, and Thomas B. Allen, *Rickover.* New York: Simon and Schuster, 1982.

Polmar, Norman, and Jurrien Noot, *Submarines of the Russian and Soviet Navies, 1718-1990.* Annapolis, Md.: Naval Institute Press, 1991.

Preston, Antony, *Sea Combat off the Falklands.* London: Willow Books, 1982.

Richelson, Jeffrey, *The U.S. Intelligence Community.* Cambridge, Mass.: Ballinger Publishing Company, 1985.

Schwab, Ernest Louis, *Undersea Warriors.* New York: Crescent Books, 1991.

Science and Technology Illustrated: The World around Us. Encyclopaedia Brittanica, 1984.

Speed, Keith, *Sea Change.* Bath, England: Ashgrove Press, 1982.

Stefanick, Tom, *Strategic Antisubmarine Warfare and Naval Strategy.* Lexington, Mass.: Lexington Books, 1987.

Submarines: Hunter/Killers & Boomers. New York: Beekman House, 1990.

The Sunday Times of London Insight Team, *War in the Falklands.* New York: Harper & Row, 1982.

Woodward, Bob, *VEIL: The Secret Wars of the CIA, 1981-1987.* New York: Simon and Schuster, 1987.

PERIODICALS

Alder, Konrad, "The U.S. Navy's BGM-109 Tomahawk Cruise Missile." *Armada International,* May 1989.

Alpern, David M., and Kim Willenson, "The Case of the Crippled Sub." *Newsweek,* November 16, 1983.

Atkeson, Edward B., "Fighting Subs Under the Ice." *Proceedings* (U.S. Naval Institute), September 1987.

Benjamin, Daniel, "Danger! Soviet Subs at Work." *Time,* July 10, 1989.

Bock, Gordon, "Run Silent, Run to Moscow." *Time,* June 29, 1987.

Booth, William, "Quiet Soviet Subs Prompt Concern." *News & Comment,* March 31, 1989.

Boyes, Jon, and W. J. Ruhe, "The Submarine SLCM Problem." *Submarine Review,* January 1990.

Bray, Jeffrey K., "Bottom Mines for Submarines." *Submarine Review,* January 1988.

Browne, Malcolm W., "Slippery Skins for Speedier Subs." *Discover,* April 1984.

Bushnell, D. M., and K. J. Moore, "Drag Reduction in Nature." *Annual Review of Fluid Mechanics,* 1991.

Byron, John L.:
"No Quarters for Their Boomers." *Proceedings* (U.S. Naval Institute), April 1989.
"The Victim's View of ASW." *Proceedings* (U.S. Naval Institute), April 1982.

Cann, Gerald A., "New Submarine Concepts." *Submarine Review,* January 1988.

Carley, William M., "How Secret Soviet Sub and Its Nuclear Arms Sank North of Norway." *The Wall Street Journal,* March 14, 1990.

Castro, Janice, "Betraying Navy—and Country." *Time,* June 3, 1985.

Chatham, Ralph E.:
"Fighting Submarines: Confuse the Bastard." *Proceedings* (U.S. Naval Institute), September 1990.
"A Quiet Revolution." *Proceedings* (U.S. Naval Institute), January 1984.

Cohen, Philip M., "Bathymetric Navigation." *Proceedings* (U.S. Naval Institute), vol. 90, no. 10.

Compton-Hall, P. R., ed., *The Submariner's World.* Vol. 1. Dobbs Ferry, N.Y.: Sheridan House, 1983.

Dantzler, H. Lee, Jr., "A Perspective of Soviet Strategic Submarine Bastions." *Submarine Review,* January 1991.

"Dead in the Water." *Time,* November 14, 1983.

DeVoss, David, "Red Hunt." *Los Angeles Times Magazine,* April 12, 1987.

Dick, Richard, "The Loss of the *Komsomolets.*" *Proceedings* (U.S. Naval Institute), October 1990.

Dobbs, Michael, "Soviet Sub Limps Home after Nuclear Shutdown." *Washington Post,* June 27, 1989.

Drew, Christopher, Michael L. Millenson, and Robert Becker:
"A Cold War Fought in the Deep." *Chicago Tribune,* January 6, 1991.

"For U.S. and Soviets, An Intricate Undersea Minuet." *Chicago Tribune*, January 8, 1991.

"A Risky Game of Cloak-and-Dagger—Under the Sea." *Chicago Tribune*, January 7, 1991.

Dworetzky, Tom, "Run Silent, Run Deadly." *Discover*, December 1987.

Fawcette, James, "Cruise Missile Technology." *Microwave Systems News*, September 1977.

Finney, John W., "Rickover, Father of Nuclear Navy, Dies at 86." *New York Times*, July 8, 1986.

Friedman, Norman, "Advanced Torpedoes." *Armada International*, May 1983.

Frigge, William J., "Winning Battle Group ASW." *Proceedings* (U.S. Naval Institute), October 1987.

Galatowitsch, Sheila, "Undersea Mines Grow Smarter and Deadlier." *Defense Electronics*, March 1991.

Gillespie, R. Doyle, "Give Me An 'N'!" *Proceedings* (U.S. Naval Institute), February 1990.

Graveson, George L., Jr., "Tactical Submarine Weapons Require Testing." *Proceedings* (U.S. Naval Institute), January 1991.

Hemond, Harold C., "The Flip Side of Rickover." *Proceedings* (U.S. Naval Institute), July 1989.

Heppenheimer, T. A., "The Real-Life Search for Red October." *Science Digest*, April 1986.

Hiatt, Fred:
"Experts Report Soviet Sub May Have Hit U.S. Gear." *Washington Post*, November 5, 1983.
"Soviet A-Sub Disabled in Atlantic." *Washington Post*, November 4, 1983.

Holzer, Robert, "Navy Program to Improve Effectiveness of Current Torpedoes." *Defense News*, March 19, 1990.

Ishchenko, P., "The Final Hours of the 'Mike.' " *Submarine Review*, October 1989.

Kaul, Ravi, "The Indo-Pakistani War and the Changing Balance of Power in the Indian Ocean." *U.S. Naval Institute Proceedings, Naval Review 1973*.

Kaylor, Robert, "The Navy's 21st-Century Submarine." *U.S. News & World Report*, April 24, 1989.

Lemkin, Bruce, "The New Leader of the Pack." *Proceedings* (U.S. Naval Institute), June 1991.

Longworth, Brian R., "Torpedo Propulsion: Then, Now, Tomorrow." *Submarine Review*, April 1989.

McCormick, Gordon H., "Stranger than Fiction: Soviet Submarine Operations in Swedish Waters." *Conflict*, 1990.

McDonnell, David C., "Strategic Thought for Submarines." *Submarine Review*, July 1991.

Magnuson, Ed, " 'Very Serious Losses.' " *Time*, June 17, 1985.

Manthorpe, H. J., Jr., "The Soviet View." *Proceedings* (U.S. Naval Institute), October 1990.

Matthews, William, "U.S. Slow to Accept New Sub Propulsion Idea." *Navy Times*, July 27, 1987.

Moore, K. J., "Emerging Technologies for Submarines." *Submarine Review*, October 1989.

Mueller, J. B., "Team Hunting: It *Can* Work." *Proceedings* (U.S. Naval Institute), October 1987.

Newton, George:
"Arctic Submarine Warfare." *Submarine Review*, October 1989.
"Factors in Arctic Submarine Ops." *Submarine Review*, July 1987.

Nylen, Daniel I., "Melee Warfare." *Proceedings* (U.S. Naval Institute), October 1987.

"An Odd Little War Turns Very Ugly." *Newsweek*, May 17, 1982.

O'Neil, Paul, "Beneath North Pole on a Well-Planned Adventure." *Life*, September 1, 1958.

Painter, Floyd C., "The SSN-21 Seawolf Attack Submarine." *Defense Electronics*, July 1988.

Pariseau, Richard, "How Silent the Silent Service?" *Proceedings* (U.S. Naval Institute), July 1983.

Pariseau, Richard R., and Lee F. Gunn, "What Quieting Means to the Soviets." *Proceedings* (U.S. Naval Institute), April 1989.

Payne, Henry E., III, "Submarine Maneuvering Instability." *Submarine Review*, January 1988.

Peppe, P. Kevin:
"Acoustic Showdown for the SSNs." *Proceedings* (U.S. Naval Institute), July 1987.
"Beyond Seawolf." *Proceedings* (U.S. Naval Institute), April 1991.

Pocalyko, Michael, "Sinking Soviet SSBNs." *Proceedings* (U.S. Naval Institute), October 1987.

Polmar, Norman, and Ray Robinson, "What Lurks in the Soviet Navy?" *Proceedings* (U.S. Naval Institute), February 1990.

Preston, Antony, "Developments in Torpedoes." *Asian Defense Journal*, March 1988.

Prina, L. Edgar:
"ASW: Duel under the Sea." *Sea Power*, July 1989.
"A Sea Full of Dragons." *Sea Power*, October 1983.

Reed, Fred, "Dive, He Said." *Harper's*, September 1988.

Roseborough, W. D., Jr., "Evolution of Modern U.S. Submarines from End of World War II to 1964." *Naval Engineers Journal*, November 1988.

Rouarch, Claude, "The Naval Mine." *International Defense Review*, September 1984.

Ruhe, W. J., "The Loss of the Soviets' Mike." *Submarine Review*, April 1991.

Russell, George, "You Must Go Home Again." *Time*, November 16, 1981.

Sandza, Richard, "Strong and Silent." *Newsweek*, September 12, 1988.

Scheina, Robert L., "Where Were Those Argentine Subs?" *Proceedings* (U.S. Naval Institute), March 1984.

Schwartz, John, "The Mind of a Missile." *Newsweek*, February 18, 1991.

"Soviet Sub Blamed in Collision." *New York Times*, March 15, 1977.

"Soviet Sub Collides with U.S. Navy Ship off Coast of Greece." *Washington Post*, August 31, 1976.

"Soviet Submarine, U.S. Frigate Collide." *Aviation Week & Space Technology*, March 14, 1977.

Stavridis, James, "Creating ASW Killing Zones." *Proceedings* (U.S. Naval Institute), October 1987.

Stefanick, Tom:
"Nonacoustic Means of Submarine Detection." *Submarine Review*, January 1990.
"Submarine Use of the Ocean Environment." *Submarine Review*, April 1989.

"Submarine Lessons of the Falklands War." *Submarine Review*, April 1983.

Tierney, John, "The Invisible Force." *Science 83*, November 1983.

Truver, Scott C., "Weapons That Wait . . . and Wait. . . ." *Proceedings* (U.S. Naval Institute), February 1988.

Tsipis, Kosta, "Cruise Missiles." *Scientific American*, February 1977.

"U.S. Navy Sees Disabled Soviet Sub Towed." *Washington Post*, November 6, 1983.

Wallace, Robert, "A Deluge of Honors for an Exasperating Admiral." *Life*, September 8, 1958.

Watts, Robert B., "Defending Our Shores." *Proceedings* (U.S. Naval Institute), October 1987.

Whitaker, Mark, "Whisky on the Rocks." *Newsweek*, November 9, 1981.

Witt, Mike, "In-Depth Communications." *Asian Defence Journal*, February 1990.

Worth, Richard A., "Defending the 100-Fathom Curve." *Proceedings* (U.S. Naval Institute), October 1987.

OTHER SOURCES

Amundsen, Kirsten, "Soviet 'New Thinking' the Northern Strategy and Special Operations in Sweden: A Study in Contrasts." Working Papers in International Studies, The Hoover Institution, Stanford University, April 1989.

"AN/BRT-1 Sub-Launched One-Way Transmitter (SLOT) Buoy." Sippican, Inc., company brochure.

Hansen, Lynn M., *Soviet Navy Spetsnaz Operations on the Northern Flank: Implications for the Defense of Western Europe*. Stratech Studies SS84-2, Center for Strategic Technology, Texas A&M University, April 1984.

"Harpoon Anti-Ship Missile." McDonnell Douglas Missile Systems Company brochure, July 1990.

Puche, Ricardo Albert, *The Malvinas War From the Argentinian Viewpoint*. Air War College Research Report, Maxwell AFB, Ala., 1988.

Index

Picture Credits

The sources for the illustrations that appear in this book are listed below. Credits from left to right are separated by semicolons; from top to bottom they are separated by dashes.

Cover: U.S. Navy Recruiting. 6, 7: Associated Press, London. 13: Imperial War Museum, London (2). 17-19: Art by Steven R. Wagner. 20, 21: Art by Steven R. Wagner—Yogi Kaufman/Yogi, Inc.; Mohammed Gad-El-Hak/Quest Integrated, Kent, Wash. (3). 22, 23: Defense Mapping Agency Hydrographic/Topographic Center, Washington, D.C., copied by Evan Sheppard; art by Steven R. Wagner. 24, 25: Fil Hunter. 26: Leo Blanchette/General Dynamics/Electric Boat Division, Groton, Conn. 29: Department of Defense, DD-ST-89-11764. 32: From *Submarine Design and Development* by Norman Friedman. © 1984 Norman Friedman permission granted by U.S. Naval Institute. 36, 37: Carl Mydans for *Life*. 38: Courtesy of U.S. Naval Institute, USN-686222. 42: Michael D. P. Flynn/Department of Defense. 45: Private Collection. 46, 47: Dagbladet/Sipa Press. 50, 51: Yogi Kaufman/Yogi, Inc. 53: Marine Nationale, Paris. 54: Department of Defense, DN-SN-87-04298. 56, 57: Department of Defense, DN-SC-90-05225. 59: John H. Sheally II/*The Virginian-Pilot/The Ledger-Star*, Norfolk, Va. 62: Mainichi Shimbun, Tokyo. 67-71: Art by Alfred T. Kamajian. 72: General Dynamics, San Diego, Calif. 76, 77: AP/Wide World. 78, 79: Jan Collsiöö/Pressens Bild, Stockholm. 82, 83: Department of Defense, K115657—K136657. 84, 85: Department of Defense, DN-ST-87-05071. 87: U.S. Navy, courtesy Yogi, Inc. 90, 91: Department of Defense. 92, 93: Steve Northup/Black Star. 94, 95: Yogi Kaufman/Yogi, Inc. 96, 97: British Crown Copyright 1991/MOD. 98: McDonnell Douglas, St. Louis, Mo. 104-105: Department of Defense (2). 106, 107: Department of Defense. 110: Bernie Campoli/U.S. Navy, courtesy Yogi, Inc. 114, 115: U.S. Navy Recruiting. 118-121: Steve Kaufman/Yogi, Inc. 122, 123: Yogi Kaufman/Yogi, Inc. 126: Department of Defense. 128-132: British Crown Copyright 1991/MOD. 138, 139: Chuck Mussi/U.S. Navy, courtesy Yogi, Inc. 141: Michael D. P. Flynn/Department of Defense. 142, 143: Courtesy of Marconi Underwater Systems Limited, Waterlooville, Hampshire, England. 144-153: Art by Steven R. Wagner.

TIME LIFE ® BOOKS

Time-Life Books
is a division of Time Life Inc.,
a wholly owned subsidiary of
THE TIME INC. BOOK COMPANY

TIME-LIFE BOOKS

PRESIDENT: Mary N. Davis

Managing Editor: Thomas H. Flaherty
Director of Editorial Resources: Elise D. Ritter-Clough
Director of Photography and Research:
John Conrad Weiser
Editorial Board: Dale M. Brown, Roberta Conlan, Laura Foreman, Lee Hassig, Jim Hicks, Blaine Marshall, Rita Thievon Mullin, Henry Woodhead
Assistant Director of Editorial Resources/Training Manager:
Norma E. Shaw

PUBLISHER: Robert H. Smith

Associate Publisher: Ann M. Mirabito
Editorial Director: Russell B. Adams, Jr.
Marketing Director: Anne C. Everhart
Production Manager: Prudence G. Harris
Supervisor of Quality Control: James King

Editorial Operations
Production: Celia Beattie
Library: Louise D. Forstall
Computer Composition: Deborah G. Tait (Manager), Monika D. Thayer, Janet Barnes Syring, Lillian Daniels
Interactive Media Specialist: Patti H. Cass

Correspondents: Elisabeth Kraemer-Singh (Bonn); Christine Hinze (London); Christina Lieberman (New York); Maria Vincenza Aloisi (Paris); Ann Natanson (Rome); Dick Berry (Tokyo). Valuable assistance was also provided by Ian Katz (Buenos Aires); Juan P. Sosa, Assel Surina (Moscow); Farhan Bokhari, Meenakshi Ganguly, and Deepak Puri (New Delhi); Elizabeth Brown, Katheryn White (New York); Dag Christensen (Oslo); Leonora Dodsworth (Rome); Mary Johnson (Stockholm).

THE NEW FACE OF WAR

SERIES EDITOR: Lee Hassig
Series Administrators: Judith W. Shanks, Myrna Traylor-Herndon

Editorial Staff for *Hunters of the Deep*
Art Directors: Christopher M. Register, Fatima Taylor
Picture Editor: Charlotte Marine Fullerton
Text Editor: Stephen G. Hyslop
Senior Writer: James M. Lynch
Writer: Charles J. Hagner
Associate Editors/Research: Susan M. Klemens, Mark G. Lazen
Assistant Editors/Research: Jennifer L. Pearce, Mark Rogers
Assistant Art Directors: Brook Mowrey, Sue Ellen Pratt
Senior Copy Coordinator: Elizabeth Graham
Picture Coordinator: David Beard
Editorial Assistant: Kathleen S. Walton

Special Contributors: Champ Clark, George Constable, Ken Croswell, George Daniels, Barbara Mallen, Anthony K. Pordes, Craig Roberts, Diane Ullius (text); Doug Brown, John Leigh, Sheila K. Lenihan, Eugenia Scharf, Barbara Jones Smith, Christine B. Soares, Joann S. Stern, Hattie Wicks, Kathy Wismar (research); Mel Ingber (index).

Library of Congress Cataloging in Publication Data
Hunters of the deep/by the editors of Time-Life Books.
 p. cm. (The New face of war).
 Includes bibliographical references and index.
 ISBN 0-8094-8637-7
 1. Submarine warfare. I. Time-Life Books. II. Series.
V210.H87 1992
359'.9'3—dc20 91-19863 CIP
ISBN 0-8094-8638-5 (lib. bdg.)